The Five Biggest Unsolved Problems in Science

Arthur W. Wiggins
Charles M. Wynn

With Cartoon Commentary
by Sidney Harris

WILEY
John Wiley & Sons, Inc.

This book is printed on acid-free paper. ♾

Published by John Wiley & Sons, Inc., Hoboken, New Jersey
Published simultaneously in Canada

ISBN 0-471-26808-9

Printed in the United States of America

10 9 8 7 6 5 4 3 2 1

Contents

Preface v

1 Science in Perspective 1

2 Physics: Why Do Some Particles Have Mass while Others Have None? 13

3 Chemistry: By What Series of Chemical Reactions Did Atoms Form the First Living Things? 37

4 Biology: What Is the Complete Structure and Function of the Proteome? 71

5 Geology: Is Accurate Long-range Weather Forecasting Possible? 101

6 Astronomy: Why Is the Universe Expanding Faster and Faster? 127

Problem Folders 159
Idea Folders 171
Resources for Digging Deeper 217
Photo Credits 224
Index 225

Preface

Here we are, human beings, situated on a chunk of rock called a planet orbiting a nuclear fusion reactor called a star that is one of a huge group of stars called a galaxy, in turn part of clusters of galaxies that make up the universe. Although our condition, which we call life, is shared by plenty of other organisms on this planet, we alone seem to have the mental equipment to seek and arrive at a general understanding of the universe and its contents. Our efforts to comprehend the nature of the universe are collectively called science. These understandings haven't been easy to achieve and are far from complete. However, we do seem to be making progress.

This book is the third of a trilogy that deals with understanding the universe. In our first book, *The Five Biggest Ideas in Science,* we examined fundamental ideas in which scientists have a great deal of confidence because of experimental evidence. Our second book, *Quantum Leaps in the Wrong Direction: Where Real Science Ends . . . and Pseudoscience Begins,* examined ideas in which scientists have little or no confidence because experimental evidence is lacking. This latest book is our effort to tell you about the biggest unsolved problems scientists are working on. Here, although there is a great deal of experimental evidence, even more is required, because no single hypothesis about each of the problems can be supported adequately. We'll look at the events and understandings that led to these unsolved problems and then bring you up to date on science's cutting-edge efforts to solve them. Sidney

Harris, America's premier science cartoonist, enlivens the discussions with his own brand of humor, which not only illustrates the ideas but illuminates them from a fresh perspective.

These unsolved problems were chosen, one from each major branch of natural science, on the basis of their explanatory power, difficulty, scope, and far-reaching implications. In addition to discussing the biggest unsolved problems, we have included a section called Problem Folders, a brief look at a selection of the other problems from each field. Any of these problems could increase in importance as more is learned about it. Also, we have included Idea Folders, which contain additional details about the background of some of the unsolved problems. Finally, there is a section called Resources for Digging Deeper, in which sources of information are listed to help you learn more about topics you find particularly appealing.

Special thanks go to Wiley senior editor Kate Bradford, who first conceived this theme, and our agent, Louise Ketz, for her timely encouragement.

Science has become like the proverbial 800-pound gorilla in our culture. The pursuit of scientific knowledge consumes enormous amounts of time, effort, and brainpower. Technological applications of scientific knowledge require correspondingly huge resources with gigantic global industries generated in the process. The thing about 800-pound gorillas is, you've got to watch them closely. We hope this book helps clarify where science is headed, so we can all keep a watchful eye on our gorilla.

AWW, CMW, SH

CHAPTER ONE

Science in Perspective

> It is the mark of an educated mind to rest satisfied with the degree of precision which the nature of the subject admits and not to seek exactness where only an approximation is possible.
>
> —*Aristotle*

Science ≠ Technology

Science and technology are pretty much the same thing, aren't they?

No.

Although the technology that dominates modern culture is driven by science's understandings of the universe, technology and science spring from entirely different motivations. Let's put the substantial differences between science and technology into perspective. While science is practiced primarily because of the fundamental desire of human beings to know and understand the universe, technology is pursued because of the fundamental desire of human beings to influence the human condition. That influence may take the form of earning a living, helping others, or even exercising power over others for personal gain.

While individuals often find themselves practicing "pure" science and "applied" science at the same time, the institution of science can carry on basic research without necessarily having an eye to eventual

products. A nineteenth-century British chancellor of the exchequer, William Gladstone, remarked to Michael Faraday about his basic discoveries linking electricity and magnetism: "This is all very interesting, but what good is it?" Faraday replied, "Sir, I do not know, but some day you will tax it." About half the current wealth of developed nations comes from Faraday's connection of electricity and magnetism.

Before scientific understandings are translated into technology, additional considerations are necessary. Besides the question of what gadget *can* be designed, there's the question of what *should* be built, a question that is properly the province of the field of ethics. Ethics is part of another whole area of people's intellectual activities: the humanities. The major difference between science and the humanities is objectivity. Science strives to study the operation of the universe as objectively as possible, while the humanities have no such goal or requirement. To paraphrase Margaret Wolfe Hungerford (nineteenth-century Irish romance novelist), "Beauty [and truth and justice and fairness and . . .] is in the eye of the beholder."

Science is far from a monolithic entity. Natural sciences study our surroundings as well as people in their functional similarity to other life-forms, whereas human sciences study people's rational/emotional behavior and the institutions set up by people for social, political and, economic interactions. Figure 1.1 is a graphical representation of these relationships.

While this neat characterization is helpful in understanding overall relationships, the real world is considerably more complex. Ethics helps

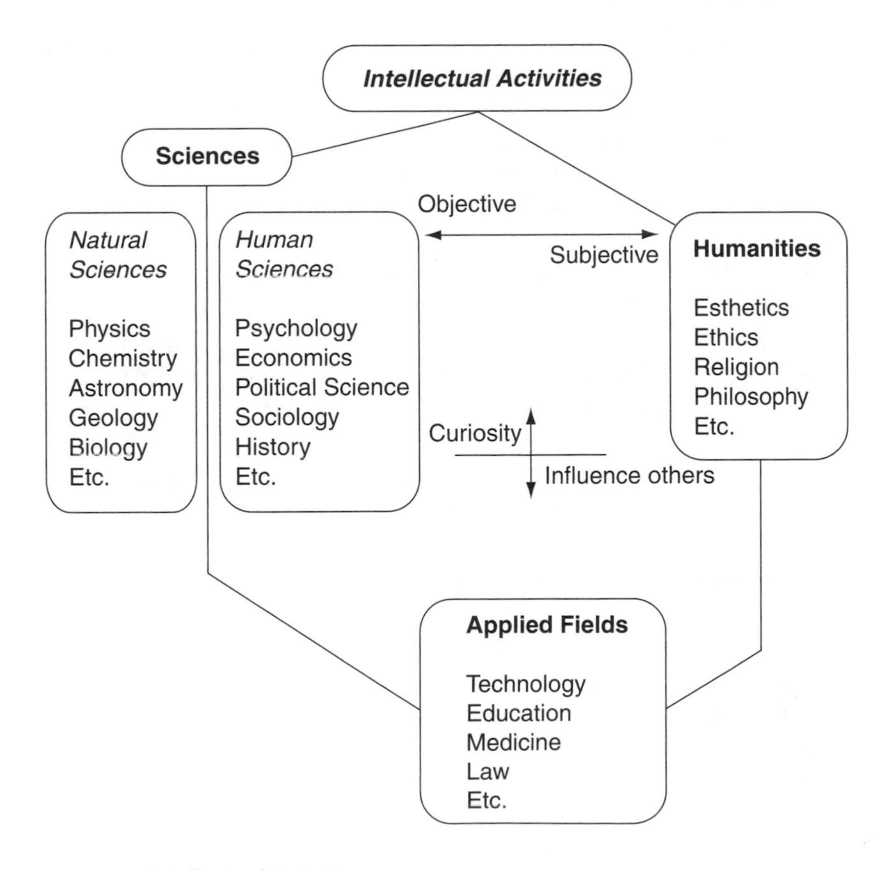

FIGURE 1.1. Intellectual Activities

dictate what topics are researched, what research methods are used, and what applications are prohibited because they are deemed potentially too dangerous to human welfare. Economics and political science also play major roles because science can only study what the culture is willing to support in terms of capital equipment, personnel, and political acceptability.

Science's Operating Procedure

The success of science in analyzing the workings of the universe is a result of the dynamic interplay between observations and ideas. This interactive process is known as the *scientific method.* (See Figure 1.2.)

During the *observation* step, some specific occurrence is perceived by the human senses with or without the aid of instrumentation. While the natural sciences have a large number of identical subjects to observe (think carbon atoms), the human sciences have a smaller number of distinctly different subjects (think human beings, even identical twins).

Human thought processes being what they are, data will be collected for just so long before the mind, in its search for order, begins to construct patterns or explanations. This is called the *hypothesis* step. The logic that uses specific observations to construct a general hypothesis is inductive reasoning. It involves making generalizations and is therefore the most precarious type of reasoning. While some people make an art form of jumping to

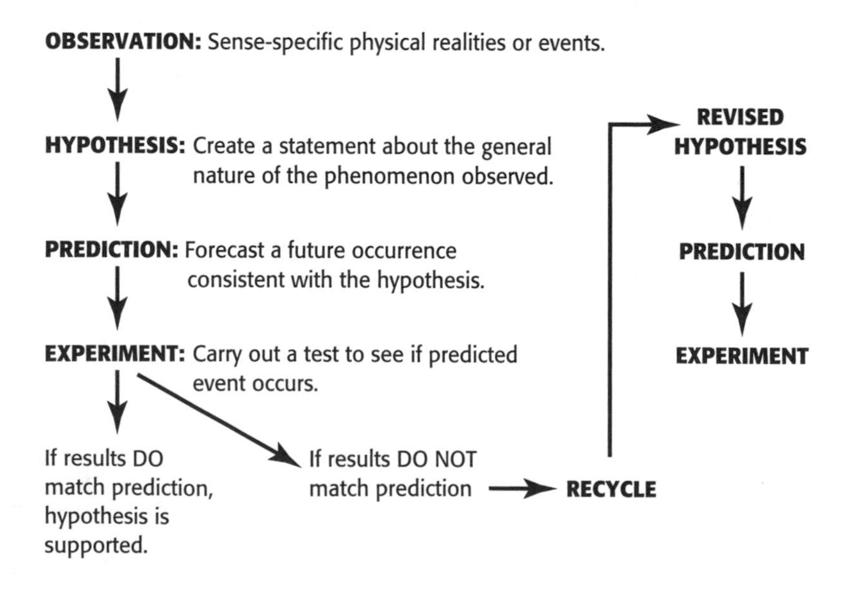

FIGURE 1.2. Scientific Method Overview

conclusions, within the context of the scientific method, such activity is restricted because succeeding steps bring the hypothesis back to reality.

Often the hypothesis is framed in whole or in part in a different language from that used in everyday speech. The language used is mathematics. Because mathematical skills require a great deal of effort to acquire, explaining scientific hypotheses to people not trained in mathematics requires translation of mathematical concepts into conversational language. Unfortunately, the meaning of the hypothesis may suffer in the process.

Once a hypothesis is formed, it can be used to forecast some future event that is expected to occur in a particular way if the hypothesis is true. This *prediction* can be derived from the hypothesis using deductive reasoning. For example, Newton's second law says $F = ma$. So, if $m = 3$ units and $a = 5$ units, then F should be 15 units. Carrying out this step is an appropriate task for computers, which operate on the basis of deductive reasoning.

After the prediction is made, the next step is to perform an *experiment* to see if the prediction is supported by evidence. Some experiments may be easy to design, but in many cases they are extremely hard to carry out. While intricate and expensive labor-intensive scientific instruments that generate much valuable data have been constructed, it is often difficult to obtain funding and then to invest the effort and patience needed to make sense of the huge amount of information obtained. Natural sciences have the advantage of being able to isolate the object of their study (think test tubes), while human sciences often have to contend with numerous variables simultaneously filtered through the minds of different people having individual agendas (think surveys).

Once the experiment phase is completed, the result is compared with the prediction. Since the hypothesis is general and the experimental results are specific instances, a result in which the experiment matches the prediction doesn't prove the hypothesis, it merely supports it. On the other hand, if the experimental result doesn't match the prediction, some aspect of the hypothesis must be false. This feature of the scientific method, called falsifiability, places a stringent requirement on hypotheses. As Albert Einstein said, "No amount of experimentation can prove me right, one experiment can prove me wrong."

A hypothesis that is shown to be false in some way must be recycled—that is, it must be modified slightly, changed radically, or abandoned altogether. The judgment about how much change is appropriate can be an extremely difficult call. Recycled hypotheses will have to work their way through the sequence again and again and either survive or fail subsequent prediction/experiment comparisons.

Another facet of the scientific method that keeps the process on target is *replication.* Any observer suitably trained and equipped should be able to repeat prior experiments or predictions and obtain comparable results. In other words, constant rechecking occurs in science. For example, a team of scientists at Berkeley Lab in California attempted to synthesize a new element by bombarding lead targets with an intense beam of krypton ions and analyzing the resulting products. The Berkeley scientists announced the synthesis of element 118 in 1999.

Synthesis of a new element is important news because of the element's novelty. In this case, its synthesis would also support previous ideas about the stability of heavy elements. Scientists at other laboratories (GSI in Germany, GANIL in France, and RIKEN Lab in Japan), however, were unable to duplicate the reported synthesis of element 118. An augmented Berkeley Lab team repeated the experiment. It, too, failed to reproduce the earlier reported results. The Berkeley team reanalyzed the original experimental data using revised software codes and were unable to confirm the existence of element 118. It retracted its claim. This refining process indicates that science's quest to understand the universe is, and must be, never-ending.

Sometimes predictions as well as experiments are rechecked. In February 2001, Brookhaven National Laboratory in New York reported an experimental result for a property known as the magnetic moment of the muon (a negatively charged particle similar to the electron, but considerably more massive) that was slightly larger than the prediction from the Standard Model of particle physics (more about this model in chapter 2). Because the Standard Model's prediction had been matched by experimental results to an extremely close tolerance for many other particle properties, this discrepancy in the magnetic moment of the muon strongly implied that the Standard Model was flawed.

The prediction of the magnetic moment of the muon was the result of a complex and lengthy calculation carried out independently by groups in Japan and New York in 1995. In November 2001, these calculations were repeated by physicists in France. The French physicists discovered an erroneous minus sign on one of the terms and posted their results on the World Wide Web. As a result, the Brookhaven group rechecked its own calculations, acknowledged the mistake, and published corrected results. The net effect of this correction was to reduce the disagreement between the prediction and the experiment. The Standard Model awaits, and must withstand, future challenges as science's never-ending search continues.

The Scientific Method in Action

Let's take a look at a classic example of the scientific method at work on a step-by-step basis.

OBSERVATION J. J. Thomson, the director of the Cavendish Laboratories in England just before the turn of the twentieth century, observed a beam of light in a cathode ray tube (forerunner of the modern TV picture tube). Since the beam (1) deflected toward positively charged electrical plates and (2) hit its target, producing individual flashes of light, it had to consist of negatively charged particles, which were called *electrons* by nineteenth-century Irish physicist George FitzGerald in his comments on Thomson's experiment. (The name *electron* had been proposed earlier as a unit of electrical charge by another Irish physicist, George Stoney.)

HYPOTHESIS Since atoms are uncharged (neutral), and Thomson had found negatively charged particles within them, he deduced that there must be some positive charge in atoms as well. Thomson theorized in 1903 that the positive charge was smeared throughout the whole atom, with the negatively charged electrons embedded inside the positive material. This depiction resembled a traditional British dessert and was therefore referred to as the Thomson Plum Pudding Model of the Atom.

PREDICTION Ernest Rutherford was an expert on positively charged particles known as alpha particles. At the beginning of the twentieth century, he predicted that if these particles were shot at atoms consisting of the sparse and smeared-out positive charge of the Thomson Plum Pudding Model, it would be like shooting pool balls at fog. Most would rip right through; very few would be deflected even slightly.

EXPERIMENT In 1909, Hans Geiger and Ernest Marsden set up an apparatus to shoot alpha particles at a thin sheet of gold atoms. The results were quite different from what they expected. Some alpha particles were deflected at large angles, and some even bounced back. Rutherford said, "It was almost as incredible as if you fired a 15-inch shell at a piece of tissue paper and it came back and hit you."

RECYCLE The Thomson Plum Pudding Model of the Atom was replaced by the Rutherford Solar System Model, in which the positive charge was concentrated in a relatively tiny nucleus at the center of the atom and the electrons (analogous to planets) moved in circular orbits around the nucleus (analogous to the Sun). Later in the twentieth century, as a result of subsequent prediction and experiment sequences, the Rutherford Solar System Model of the Atom was replaced by other models. Whenever experimental evidence doesn't match the prediction of an existing hypothesis, it's time to recycle the hypothesis.

Similarly, Isaac Newton's motion analysis and James Clerk Maxwell's electricity and magnetism classic hypotheses were interpreted to mean that space and time were absolute—an attractive notion. Einstein's Special Theory of Relativity replaced these comfortable absolutes with counterintuitive and philosophically unsatisfying relative quantities. The main reason relativity was accepted was that its prediction matched experimental evidence.

In spite of the popularity of an earlier idea, the celebrity status of a theory's proponents, the unattractiveness of a new theory, the political views of an idea's author, or the difficulty in understanding the idea, the bottom line is: *Experimental evidence rules.*

Complications

The scientific method we've presented here is a rational reconstruction of the way science actually works. This idealization of the process is neater than the one that occurs in the day-to-day world. Many people may be involved and lengthy periods of time may elapse between steps that don't occur sequentially. Nevertheless, the opportunity to look back over science's development affords us the luxury of 20/20 hindsight.

A number of complicating factors must be considered. First of all, science makes several philosophical presuppositions with which some philosophers disagree. Science presumes the existence of an objective reality independent of the human observer. Without such objectivity, otherwise identical observations and experiments repeated in various labs could differ, and it would be impossible for researchers to come to a mutually agreed on hypothesis. Further, science presumes that the universe is and has always been governed by a set of fixed laws, and that these laws are ones humans are capable of understanding. If the universe's governing principles were without pattern, or if we couldn't make sense of them, no hypotheses would emerge from science's efforts. Since our understanding of these laws seems to be growing, and predictions based on them are supported by experiments, these presumptions seem reasonable.

Because science's hypotheses deal with events occurring over a broad span of time, many deal with past events that cannot be directly checked by experiment. The usual solution to this problem is to cross-check hypotheses from several sciences, seeking mutual agreement. For example, the more than 4-billion-year age of Earth is supported by astronomers' measurement of helium abundance in the Sun, geologists' measurement of plate movements, and biologists' measurement of coral growth.

Especially because experimental results are unavailable for some phenomena (for example, from the distant past when there were no human observers or from an inaccessible part of the universe), more than

"WELL, I'VE BEEN THINKING HARDER THAN YOU HAVE, AND MY THOUGHT EXPERIMENT DISPROVES YOUR THOUGHT EXPERIMENT."

one hypothesis can be advanced to explain some event. The ticklish situation of having multiple hypotheses coupled with no possibility of experimental resolution is dealt with by a principle of scientific economy referred to as Ockham's Razor. The English philosopher William of Ockham (1285–1349) was a Franciscan monk who often used a common medieval principle in his philosophical writings: "Plurality should not be assumed without necessity." The military has given this principle a simpler and more direct expression: KISS—Keep It Simple, Stupid; or Keep It Short and Sweet. Expressed either way, it gives guidance in the absence of experimental evidence. If several hypotheses exist, and no experiment can be performed to choose between them, choose the simplest one.

Experience has shown this course to be wise. For example, in 1971, the X-ray-measuring satellite *Uhuru* found unexpectedly strong X-ray radiation referred to as Cygnus X-1 from the constellation Cygnus (the Swan). There was no apparent source for these X rays, which turned out to emanate from seemingly empty space near the supergiant star named HDE 226868, located about 8,000 light-years from Earth. (See Idea Folder 14, Compiling Star Catalogs, for an explanation of the HDE designation.) One hypothesis to explain this result was that HDE 226868 had an invisible companion. This phantom attracted mass that spewed out of HDE 226868. As this material was drawn into the unseen companion, its temperature increased enough to emit X rays. A different hypothesis requires at least two unseen bodies interacting with HDE 226868—a

normal star too dim to be seen and a rotating neutron star (the core of a star after it has lived out its life cycle and collapsed into a neutron ball) called a pulsar. These three bodies, arranged in a particular fashion, could emit X rays similar to the ones measured.

Cygnus X-1's distance renders it inaccessible to direct testing, not to mention that all its radiation was emitted around 8,000 years ago. So, which of the competing hypotheses is justified? On the basis of experimental support, either one. Using Ockham's Razor, the simpler explanation involving only one body is deemed more likely. Thus, Cygnus X-1 became the first recorded instance of an unseen companion known as a *black hole.* Subsequently, more than 30 such objects have been found under similar circumstances.

Ockham's Razor functions only when appropriate experimental support is unavailable. Its operating principle is to choose the simplest hypothesis consistent with the observations. It cannot, however, rule out other hypotheses supported by evidence, regardless of the hypothesis's more complicated nature. It cannot overrule experimental support, either. Occam's Razor is certainly less desirable than solid experimental evidence, but sometimes it's all we've got.

Unsolved Problems

Now that you have seen how science fits into the overall scheme of human intellectual activity and how it operates, you can appreciate that its open architecture allows many different paths to an increased understanding of the universe. New observations are made. Existing hypotheses are silent about these phenomena. New hypotheses are formulated to replace the silence with effective ideas. Predictions are improved. Innovative experimental apparatus is designed. All these activities lead to hypotheses that more accurately reflect the operation of the universe. The overall objective of these activities is to make sense of the universe, from its most minute details to its broadest sweep.

Science's hypotheses can be considered answers to questions or problems about the universe. Our aim here is to explore the *five biggest problems* that are at present unsolved. By "biggest," we mean the ones that have the broadest explanatory power, are the most difficult, have the most far-reaching implications, are the most critical to our understanding, or have the most potential applications. We will limit our exploration to the one biggest unsolved problem from each of the five natural sciences and try to describe the kind of progress we can anticipate toward each one's solution. Certainly the human sciences, humanities, and applied fields have important unsolved problems (for example, the nature of consciousness), but these are beyond the scope of this book.

Here are our candidates for the *biggest unsolved problem* in each of the five natural sciences, along with our justification for their selection.

PHYSICS: Motion-related properties of masses, such as velocity, acceleration, and momentum, are well understood, as are kinetic and potential energy of masses. The nature of mass itself, which is a property of many but not all of the fundamental particles of the universe, is *not* understood. The biggest unsolved problem in physics is: Why do some particles have mass while others have none?

CHEMISTRY: Chemical reactions of both nonliving and living entities have been studied extensively, with much success. The biggest unsolved problem in chemistry is: By what series of chemical reactions did atoms form the first living things?

BIOLOGY: The genome, or molecular blueprint of many life-forms, has recently been mapped. Genomes encode information about a life-form's collective proteins, or proteome. The biggest unsolved problem in biology is: What is the complete structure and function of the proteome?

GEOLOGY: The plate tectonics model satisfactorily describes the effects of interactions between the outermost of Earth's layers. But Earth's atmospheric phenomena, most notably weather patterns, seem to defy attempts to formulate models that lead to reliable predictions. The biggest unsolved problem in geology is: Is accurate long-range weather forecasting possible?

ASTRONOMY: Although many aspects of the universe's overall structure are well known, its dynamics are less well understood. Recent discoveries that the universe's expansion rate is increasing make it likelier that the universe will expand forever. The biggest unsolved problem in astronomy is: Why is the universe expanding faster and faster?

Many other interesting questions related to these problems will arise along the way, some of which may turn out to be the biggest questions of the future. These will be discussed briefly in Idea Folders at the end of the book.

William Harvey, the seventeenth-century English physician who discovered the nature of blood's circulation, said, "All that we know is still infinitely less than all that remains unknown." Stay tuned as new questions arise faster than old ones are answered. As science's circle of light expands, so does the circumference of darkness it encounters.

CHAPTER TWO

PHYSICS

Why Do Some Particles Have Mass while Others Have None?

The baby figure of the giant mass
Of things to come at large.

—*William Shakespeare,*
Troilus and Cressida

Physics is the study of the properties of matter at rest and in motion and of various forms of energy. Motion-related properties such as velocity, acceleration, and momentum are well understood, as are kinetic and potential energy. What is *not* understood is the nature of mass, a property of most forms of matter. In fact, the origin of mass is currently the biggest unsolved problem in physics.

Mass

We are all too familiar with the fact that mass exists. Mass is self-evident. We all seem to have too much or too little of it. Mass is what makes it harder to push a stalled car than a baby carriage. Mass is what gravity tugs on to keep us grounded on planet Earth.

What is not at all clear is the *origin* of mass. Many, but not all, of the fundamental particles in the universe have mass. Why do some have mass, while others have none? What "gives" mass to some particles and not others? Among particles with mass, why do some have more mass than others? Do particles without mass lack something else besides mass? The answers to these questions may lie in something called the Higgs field, but we'll need some background information before we can make sense of the elusive Higgs field.

To begin with, we know that a body's mass is related to the amount of matter it contains, and we have a pretty good idea of what forms matter: collections and combinations of atoms. But what forms the atoms? Atoms are composed of electrons, protons, and neutrons. But what forms the electrons, protons, and neutrons? Electrons are fundamental (not composed of "anything else"), but protons and neutrons are not. They are made of quarks, and it is quarks and electrons that appear to be truly fundamental.

Before we can discuss the unsolved nature of mass, we've got to see where quarks came from. (See Figure 2.1.) In the process, we'll run across a few other fundamental particles, we'll see that fields are

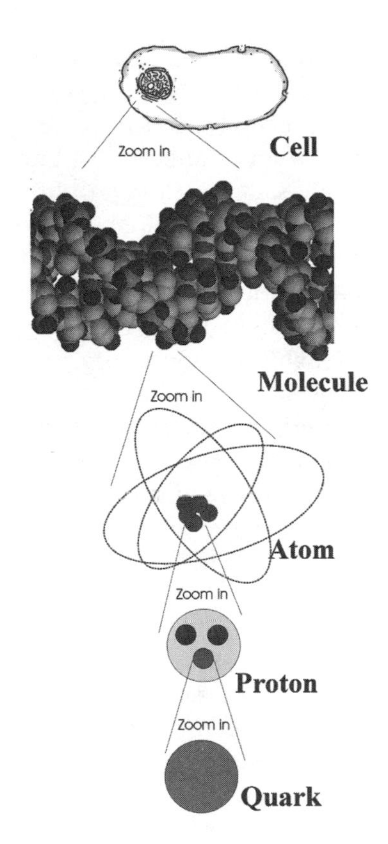

FIGURE 2.1. Building Blocks of Matter

regarded as even more fundamental than particles, and we'll investigate physics' best field theory, called the Standard Model. Next we'll note the inadequacies of the Standard Model: It doesn't identify the source of mass, and it omits gravity entirely. Finally, we'll explore theories beyond the Standard Model that would solve the problem about mass's origins.

More Fundamental Than Atoms

To find out about quarks, we must start with atoms. In the early 1900s, Ernest Rutherford's alpha particle exploration of the atom revealed the nucleus (see chapter 1), but that was only the beginning. Physicists' probes, both experimental and theoretical, have revealed a much finer level of detail about the basic building blocks of the universe. Before 1920, the atom's structure had been sorted out enough to realize that it contained a nucleus of positively charged protons and neutral neutrons (although experimental evidence for neutrons was lacking at the time), with electrons circulating around the nucleus.

The tidiness of this picture was soon shattered. To explain radiation of light from heated bodies, in 1900, German physicist Max Planck came up with the notion that light's energy only came in packets, called quanta (the singular is *quantum*), rather than in any amount, as previously thought. (Think links as opposed to bulk sausage.) To Planck, this was just a mathematical convenience that solved the problem. In 1905, however, Albert Einstein took Planck's quantum idea more seriously. He showed that if light actually has a discrete quantum nature, puzzling aspects of the photoelectric effect could be explained easily.

The photoelectric effect occurs when light is shined on a metal, causing electrons to be emitted from the metal. Electrons are emitted promptly, but if the light has too low a frequency, no electrons would be given off, no matter how intense the light source. Einstein pointed out that light acted like a particle, giving up all its energy to an electron, freeing it. Further, the Planck relation linking energy to frequency explained the lack of electron emission at low light frequency. Photons of light just didn't have enough energy to turn electrons loose. Light's behavior seemed to be much more particlelike than wavelike.

Extending the quantum idea to atoms in the 1920s produced the Quantum Mechanical Model of the Atom. This theory reversed light's situation by treating an atom's particlelike electrons as if they had wavelike properties. Quantum mechanical predictions about the colors of light given off by excited atoms matched spectroscopic evidence, so the theory passed its experimental test. The symmetry was now complete. Light could appear as a wave or a particle, and an electron (or proton or neutron) could appear as a particle or a wave, all depending on the type of experiment we perform.

One of the consequences of quantum mechanics was called the Heisenberg Uncertainty Principle. This principle says that there is a limit on the mathematical product of the position uncertainty and the momentum uncertainty of any particle and a corresponding limit on the product of the uncertainty in energy and the uncertainty in time. The principle means that the more precisely the position of an electron is determined, the less precisely its momentum can be known, and vice versa. The limit is very small and has almost no effect on measurements of objects of ordinary size. However, the philosophical consequences are enormous: *There is a limit to our knowledge.* Many scientists, including Albert Einstein, had great difficulty accepting this idea. Nevertheless, it is derived from a successful hypothesis, so accept it we must.

Next, quantum mechanics needed to be integrated with the other revolutionary idea of the early twentieth century, Einstein's Special Theory of Relativity. In 1928, British physicist P. A. M. Dirac was able to unite these theories. His new formulation was not only comprehensive, it had an interesting sidelight: It predicted the possibility of a new particle, identical to the electron except that its charge would be +1 while the electron's is −1. This particle was called an anti-electron, or positron (positive electron).

Cosmic Rays to the Rescue

Predictions about new particles are one thing, but experimental evidence was still needed. Since no one had evidence of a positron, its existence was questionable. In the early 1930s, American physicist Carl D. Anderson began to use a new probe to study matter—cosmic rays. Cosmic rays are mostly protons, alpha particles (2 protons plus 2 neutrons stuck together; in other words, a helium nucleus), or light of various frequencies. These particles exhibit a wide spectrum of energies and bombard Earth from all directions, arriving at the rate of roughly one particle per second. Although the Sun is a major source of cosmic rays, more highly energetic cosmic ray particles that are observed must originate from more energetic processes than the Sun can muster.

Cosmic rays are not visible themselves, and their effect on matter is too small to see. Anderson used two devices to detect the particles: a strong magnetic field and a Wilson cloud chamber. The magnetic field bent the paths of the charged particles inside the cloud chamber, which contained clean air saturated with water vapor. Particles zipping through the chamber ionized air molecules. The ionized air molecules provided collection points for the water vapor, eventually forming clouds. The clouds allowed tracks to be observed, much as Earth's atmosphere permits a high-flying airplane's trail to be seen even though the plane itself may be invisible to the eye.

Anderson performed the experiments in Colorado, where the high altitude minimized the shielding effects of Earth's atmosphere on incoming cosmic rays. Sure enough, one of his photographs revealed a track that bent the opposite way from the track of an electron. He had found evidence of the positron. It turns out that almost all particles have antiparticle partners, which differ in electric charge and also in more subtle ways. (See Idea Folder 1, Anti-matter.)

Because Dirac's hypothesis had predicted the existence of positrons, finding them provided support for his relativistic quantum mechanics (quantum mechanics modified by the addition of special relativity). Another Anderson discovery, however, proved much more troublesome. He found tracks for two new particles, each of which had more than 200 times the mass of the electron. One was positively charged and the other was negatively charged. They were eventually named the muon and antimuon. Their existence and properties were perplexing because muons had no clearly identifiable place in the organization of matter. Physicists' consternation was typified by the response of the Nobel Prize winner

"PARTICLES, PARTICLES, PARTICLES."

I. I. Rabi, who, upon first hearing the announcement of the discovery of the muon, said, "Who ordered that?"

The Four Forces

As if new particles weren't confusing enough, in the 1930s, new forces also were discovered. In addition to the already-known gravity and electromagnetism, two new forces were found: the *strong nuclear force,* which kept the protons and neutrons together in the nucleus, and the *weak nuclear force,* which was responsible for some nuclear radioactive decay processes. Interestingly, both strong and weak nuclear forces reached their maximum strength at very small distances and diminished to zero if the particles were any farther apart than nuclear distances. That's why we hadn't seen them—they worked over distances below our detection capability.

In the 1930s, Italian physicist Enrico Fermi formulated a theory of the weak nuclear force that predicted the existence of yet another new particle. This particle was electrically neutral and was required to carry away missing energy in one of the observed decay processes. He called it "the little neutral one," or *neutrino.* The neutrino was almost a phantom particle, interacting with ordinary matter so infrequently that it would take an eight light-year thickness of lead (more than twice the distance between our Sun and the nearest star) to stop half the neutrinos in a group. Nevertheless, neutrinos were found experimentally by American physicists Frederick Reines and Clyde Cowan (but not until 1953). Just what physics needed—another particle.

Particle Fragments, or Breaking Up Is Hard to Do

Scientists desperately needed experimental tools to explore these new particles, but the cosmic rays were too haphazard because their energies varied so widely and they came from seemingly random directions. New machines were designed in the early 1930s to carry out systematic experiments using beams of particles with known energy. The machines, called *particle accelerators,* became the chief instrument of high-energy physics, serving in the same capacity as biology's microscope and astronomy's telescope.

Two fundamentally different kinds of accelerators were designed: the linear accelerator and the circular accelerator, also called the cyclotron. In a linear accelerator, electrons are accelerated by electric fields along a long vacuum track (the Stanford University model is over two miles long) and bent by magnets to collide with a target. Detectors record the collision products.

In a cyclotron, charged particles are accelerated in the gap between the two halves of the cyclotron (called dees, because of their shape), and their paths are bent by magnetic fields within the dees. As the particles become more energetic, they travel in wider arcs. When they finally reach maximum energy, they are deflected out of the cyclotron and into the target area, where collisions occur and the impact fragments are observed by detectors. (See Idea Folder 2, Accelerators.) The cyclotron's inventor, Ernest O. Lawrence, worked tirelessly to build larger and larger cyclotrons, which he referred to as "proton merry-go-rounds," but he ran into obstacles that slowed his merry-go-rounds to a halt.

World Affairs Interrupt

The 1930s brought another upheaval: World War II.

Besides diminishing research funding, the war effort absorbed the vast majority of physicists, putting them to work on the Manhattan Project beginning in 1941. Initially, the aims of this project were to explore the energy given off by the splitting the nuclei of large atoms such as uranium, to ascertain whether this energy could be used in a weapon, and if so, to accomplish this feat before the Nazi physicists who were presumably working on a similar project. (The play *Copenhagen* by Michael Frayn explores the German/Allied nuclear bomb projects in terms of the relationship between physicists Niels Bohr and Werner Heisenberg.)

While the physics part of the task was the understanding of the nucleus, the technology part involved turning this knowledge into an explosive device. The ethics aspect came to a head when the Germans were defeated, although they never developed such a bomb. After Germany surrendered on VE day in 1945, some physicists in the United States quit the Manhattan Project. Others stayed on and developed both the A-bomb (atom bomb) and the H-bomb (hydrogen bomb), with consequences that we live with today.

Physics Resumes

After the war ended, the race to find new particles was back on, and the accelerator was the tool of choice. Particles were sent crashing into tar-

gets, and the collision fragments were analyzed carefully. At the relatively low energies available at the time, protons stuck to large nuclei, synthesizing short-lived larger nuclei. Some of these products were radioactive and decayed into smaller nuclei and other particles. The larger nuclei enlarged the Periodic Table of elements, to the delight of chemists, but physics got no new particles.

So, still bigger cyclotrons were built, producing even more energetic particles. Because of the equivalence of mass and energy (Einstein's famous $E = mc^2$), more energetic collisions have the possibility of creating more massive particles. As higher energies were achieved, the accelerators struck pay dirt. Cloud chamber pictures revealed tracks of particles that had never been seen before: charged pions (π^+, π^-) and kaons (K^+, K^-), neutral pions and kaons, the lambda particle, the sigma particle, and others, as well. Although the particles were unstable and decayed into more familiar ones after a brief existence, they were indications that matter had more surprises in store.

The particle race heated up. More cyclotrons came into operation, and their designs improved. In a device called a synchrotron, the accelerating field was synchronized to ensure a constant particle beam radius. The cloud chamber was replaced with a bubble chamber, in which the formation of bubbles in superheated liquid hydrogen made particle tracks visible. It was almost like analyzing photographs of an exploding haystack and searching for briefly existing needles. For his thesis project, a fellow graduate student of one of the authors analyzed 240,000 bubble chamber photographs, one picture at a time!

The result of all this effort was a virtual explosion of particles: well over 100 were found. Nobel laureate Enrico Fermi remarked to his student Leon Lederman (eventually a Nobel laureate in his own right), "Young man, if I could remember the names of all these particles, I would have been a botanist."

Bring on the Quarks

This expanded collection of particles created a situation in physics analogous to what chemistry faced prior to the development of the Periodic Table of elements by Dmitry Mendeleyev in 1869. There had to be some underlying structure, but what was it? Physicists devised various arrangements of the particles on theoretical grounds, searching for patterns within their organization. Heavy and medium mass particles were called hadrons and were further subdivided into baryons and mesons. All hadrons participated in the strong interaction. Less massive particles, called leptons, participated in the electromagnetic and weak interactions. But just as electrons, protons, and neutrons were needed to explain the collection of elements, something more fundamental was necessary to explain all these particles.

In 1964, American physicists Murray Gell-Mann and George Zweig, working independently, proposed a new scheme. All the hadrons could be represented as composites of three smaller particles and their corresponding anti-particles. Gell-Mann dubbed these new fundamental particles *quarks,* after the line from James Joyce's novel *Finnegan's Wake,* "Three quarks for Muster Mark." These (first) three quarks were called up (u), down (d), and strange (s) and carried fractional electric charges of $+2/3$, $-1/3$, and $-1/3$, respectively, with opposite charges for the corresponding anti-quarks.

Using this model, protons and neutrons could be built from three quarks: uud and udd, respectively. The large group of newly found mesons could be made from quark–anti-quark pairs. For example, the negative pion would consist of a down quark and an anti-up quark. The quark idea was proposed tentatively, and though it solved the problem of organizing the vast collection of particles in a mathematical sense, the reality of quarks was suspect since none had been observed—yet.

Experimentally, protons and neutrons were just fuzzy lumps, similar to the atom as depicted by the Thomson Plum Pudding Model. However, protons and neutrons were considerably smaller and couldn't be probed by shooting alpha particles at atoms as Rutherford had done. Alpha particles were too big and wouldn't reveal anything.

A Massachusetts Institute of Technology/Stanford team, working at the Stanford Linear Accelerator, investigated the nucleus by shooting electrons at hydrogen and deuterium (the heavier isotope of hydrogen whose nucleus contains one proton and one neutron). They measured the angle and energy of the electrons after the collision. At lower electron energies, the scattering was consistent with protons and neutrons being "soft" structures that would deflect electrons only slightly. But when they used electron beams of record high energy, they found that some electrons lost most of their initial energy and were scattered at large angles. With remarkable similarity to the alpha scattering work of Rutherford in the initial identification of the nucleus, American physicists Richard P. Feynman and James Bjorken interpreted the electron scattering data as indicative of an inner structure to protons and electrons—namely, the earlier-theorized quarks. Now the quark hypothesis had to be taken seriously.

Theory Strikes Back: Unification

There is a great drive in physics to simplify matters by combining theories. Near the end of the nineteenth century, James Clerk Maxwell's recognition that electricity and magnetism were simply two facets of the same phenomenon allowed the two to be united. The combination was called electromagnetism. In the 1950s, American physicists Richard P. Feynman and Julian Schwinger, and Japan's Sin-Itiro Tomonaga combined electromagnetic theory with quantum mechanics to form *quantum electrodynamics,* referred to as *QED.* In this theory, electrons interact with each other by exchanging photons of light. The photons cannot be observed because the electrons emit and absorb them within a region governed by the Heisenberg Uncertainty Principle (see page 16). Because they are not capable of being observed, they are called virtual photons.

While quarks were being hunted experimentally in the late 1960s, another theoretical unification scheme was proposed, this one involving two of the four fundamental interactions. Steven Weinberg and Sheldon Glashow in America, and Pakistani physicist Abdus Salam in Trieste, Italy, working separately, formulated a theory that unified the electromagnetic and weak interactions into one called the electroweak interaction. Besides explaining already-observed events in a more general context, this new theory added to the particle list by predicting the existence of new particles: a neutral, weakly interacting particle (now called the Z^0), the W^+ and W^-, and a massive particle called the Higgs (more about this later).

In 1973, yet another theoretical development occurred: A quantum field theory of the strong interaction was first suggested by Murray Gell-

Mann and German physicist Harald Fritszch. This theory, called *quantum chromodynamics,* was similar to QED in that basic particles, quarks, interacted by exchanging virtual (in the uncertainty zone) massless particles called gluons. Since no one had seen a gluon, one more predicted particle needed supporting evidence.

The Standard Model

By the mid-1970s, theoretical and experimental developments were digested and summarized into a single theory called the Standard Model. Mathematical reasoning beyond the scope of this book underlies many of these ideas, so be aware there is heavy-duty math lurking in the background.

Fundamental to the Standard Model is the notion that the basic building blocks of the universe are fields, not particles. Originally, fields were introduced to deal with the classic action-at-a-distance problem. How can one body influence another body when they are some distance apart and there is no matter between them? Newton's answer was that they exerted a force on each other.

To understand fields, we need to take this abstraction one level further. Remove one body. Now think of the remaining body as possibly influencing any body that happens to come by. This influence constitutes the field exerted by the body that is present. Viewed this way, *a field is a* possibility *of a force.* For example, a gravitational field is often pictured as arrows pointing toward a mass, as in Figure 2.2. There are no physical lines or arrows in space, but if an actual body is introduced at any point, it will feel a force, and the force will point in the direction shown by the arrows.

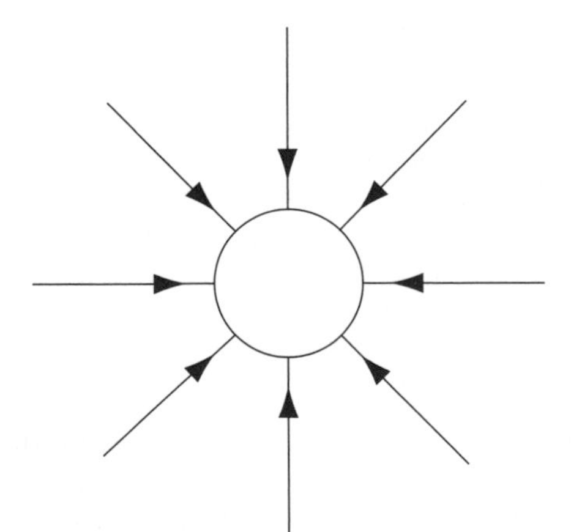

FIGURE 2.2. Gravitational Field Diagram

In similar fashion, electric field lines surround charges, and magnetic field lines surround magnets. Because iron filings are so strongly magnetic, physical models and textbook pictures that show patterns of iron filings around magnets have helped make magnetic fields seem tangible.

Although the field idea began as a conceptual tool, it now plays a central role in physics. According to the Standard Model:

- The universe's fundamental building blocks are fields.
- Tiny packets of energy (quarks or leptons) result when quantum principles are applied to fields.
- Interactions between particles are carried by the exchange of other energy packets (bosons) that cannot be observed because of uncertainty considerations.

So, the classical action-at-a-distance picture of a force between particles has been replaced by an interaction consisting of the exchange of virtual energy packets (formerly waves) between quantized bundles of field energy (formerly particles). Now, *there's* a radical departure from earlier thinking.

The Standard Model includes two interactions, the strong and the electroweak.

1. *The Strong Interaction:* The particles that result from applying quantum principles to one set of fields are called quarks. There are now known to be six quarks (and associated anti-quarks), grouped into three families, as shown in Figure 2.3. They have been named (whimsically):

   Family 1: up and down
   Family 2: charm and strange
   Family 3: top and bottom

   Quarks interact with each other by the strong interaction, which involves the exchange of virtual particles called gluons.

2. *The Electroweak Interaction:* The particles that result from applying quantum principles to another set of fields are called leptons. There are six leptons (and corresponding anti-leptons) grouped into three families, as shown in Figure 2.4. They are called:

   Family 1: the electron and electron neutrino
   Family 2: the muon and muon neutrino
   Family 3: the tau and tau neutrino

   Leptons interact by exchanging virtual particles called the photon, two W's and one Z.

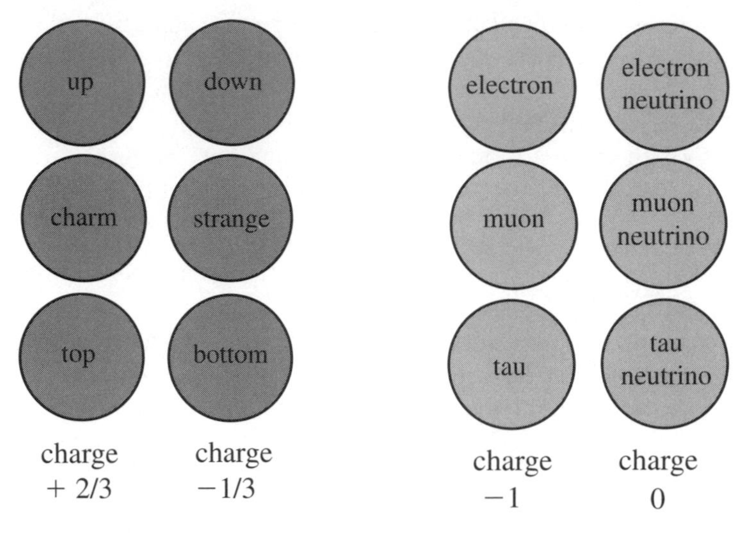

FIGURE 2.3. Quarks

FIGURE 2.4. Leptons

In summary, Figure 2.5 shows the fundamental particles and interaction carriers.

The table on page 27 lists each of the particles and their spins, charges, and masses. Notice the enormous range of masses as you scan down the mass column—more about this later.

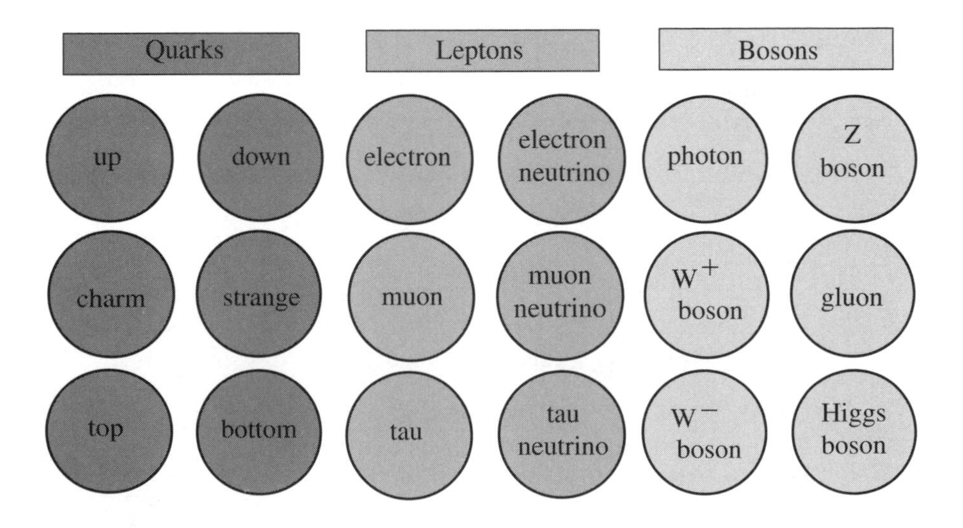

FIGURE 2.5. Fundamental Particles

Fundamental Particles and Their Masses

Particles	Approximate Mass (Units of GeV*)
Fermions	
up quark	5×10^{-3}
down quark	9×10^{-3}
electron	0.51×10^{-3}
electron neutrino	$<7.2 \times 10^{-9}$
charm quark	1.35
strange quark	0.175
muon	0.106
muon neutrino	$<2.7 \times 10^{-4}$
top quark	174
bottom quark	4.5
tau	1.78
tau neutrino	$<3 \times 10^{-2}$
Bosons	
photon	0
W^+ and W^-	80.2
Z	91.2
gluon	0
Higgs (no evidence yet)	63 to 800

* Mass is given in energy units, *GeV*, which are billions of electron volts, based on Einstein's mass-energy equivalence, $E = mc^2$.

According to the Standard Model, here's the way an atom works. Protons and neutrons are bound in the nucleus by virtual gluon exchange between the quarks that make them up. Electrons are bound to the protons in the nucleus by virtual photon exchange. Note that three families of quarks are matched exactly by three families of leptons. Of course, no one knows why there are exactly three of each. The first families of quarks and leptons are stable and constitute all the matter around us. The other two families are unstable and, after a short time, decay into their more stable cousins. If you're wondering about the possibility of more families of quarks and leptons, there have been two experimental confirmations that the number of these families is three. One is based on 1998 accelerator results for the number of neutrinos in a particular particle (lambda zero) decay, and the other rests on astronomical data (more about this in the next section).

All the particles listed have nonzero mass except for the gluon and the photon. The zero mass of the photon accounts for the long range of the electromagnetic interaction because its carrier can move at the speed of light. The weak interaction has a much shorter range because its carrier particles have substantial mass and cannot move as fast as photons. All quarks and leptons obey a set of statistical rules set up by Fermi and Dirac, and are collectively called fermions. Interaction carriers are governed by another set of rules formulated by Indian physicist Satyendranath Bose and Einstein, and they are called bosons. (See Idea Folder 3, Fermions and Bosons.)

Testing the Standard Model

The Standard Model was first named in 1974. At that time, seven particles predicted by the theory had not yet been detected. During the subsequent 20-year experimental search, all but the Higgs particle were found in successively more energetic accelerator collision experiments.

Besides observations of particles themselves, many particle properties predicted by the Standard Model have been tested experimentally, and there has been remarkable agreement between prediction and experiment. One example is the Lamb shift. In 1947, American physicist Willis Lamb measured the difference in frequency of radiation absorbed or emitted when hydrogen makes a transition between two closely spaced energy levels. Much later, using the Standard Model, the frequency of light given off by this transition was predicted to be 1057.860 +/− 0.009 MHz. Lamb originally measured the frequency to be 1057.845 +/− 0.009 MHz. The two values differ by only about one part in 100,000. When the listed uncertainties are taken into account, prediction and experiment overlap substantially. This remarkable agreement is true of many other prediction/experiment sequences involving the Standard Model, which provides the theory with extremely strong support.

Searching for more and more massive particles required larger and larger accelerators. For economic reasons, physics needed a lower-cost testing method. Just as Carl Anderson had used naturally occurring cosmic rays as high-energy particle sources, physicists searched for a natural occurrence of the much higher-energy particles theorized by the Standard Model. The only time such high-energy particles had existed naturally was during the first few moments of the Big Bang, when all matter and energy in the universe began expanding. Conditions during the first small fraction of a second of the Big Bang's primeval fireball were spectacularly hot and dense. Under these conditions, all families of fundamental particles were present, so the initial moments of the Big Bang would have provided a great laboratory for testing the Standard Model. Even though that era is inaccessible, predictions about conditions that exist now may be made and compared with measured reality.

Particle/astrophysicist David Schramm often repeated a quote from Soviet physicist Iacov Zeldovich: "The universe is the poor man's accelerator; experiments don't need to be funded, and all we have to do is collect the experimental data and interpret them properly." For example, if *four* families of fundamental particles existed, the amount of helium formed during the first few minutes of the Big Bang expansion would constitute more than 26% of the present-day universe. *Three* fundamental particle families would produce only a 25% helium abundance. Since 25% is the the amount of helium that has been detected, the Standard Model's inclusion of exactly three families of quarks and leptons has strong experimental support.

The convergence of high-energy physics and astrophysics in analyzing the first few instants of the Big Bang expansion has provided many fruitful insights. For example, combining three basic universal constants (Planck's constant, the speed of light, and the gravitational force strength constant) in several ways yields minimum values for fundamental quantities such as time, mass, and energy. These are called the Planck scales:

time: 10^{-43} sec   length: 10^{-35} m   energy: 10^{9} Joules

If the Planck energy is concentrated in a volume the size of the Planck length cubed, the equivalent mass ($E = mc^2$) contained in this small space would be so dense that light couldn't escape and would be cut off from the rest of the universe—it would be a black hole. Viewed in this way, no smaller distances or times than the Planck scale values have any meaning, so this is the level below which space and time are simply "quantum foam," where all the laws of physics no longer hold.

Reasoning based on the Planck scales provides a possible picture of the universe's beginning. A submicroscopic quantum fluctuation underwent an inflationary phase of enormously rapid expansion, in which transitions occurred and the four forces "froze out" as the temperature cooled, analogous to the way liquid water turns to ice. If interactions with the Higgs field determine particle masses (see page 31), these masses might have become fixed at random values, depending on the details of the cooling process. If that is the case, many universes might have formed in similar fashion, but with slightly different values for the masses of the fundamental particles. See page 171 for an additional discussion of this topic.

The Standard Model's Dark Side

There are several classes of objections to the Standard Model. The first objection is on mathematical grounds. Solving the Standard Model's equations for particle properties often involves an approximate mathematical technique called perturbation. A quantity's value is determined to whatever precision desired by including more and more terms in a mathematically determined series, based on successively larger powers of a quantity called a parameter. As long as the parameter is small, successive terms in the series become smaller, so not very many terms may be needed to obtain the required accuracy. But because not all parameters are small, some calculations require many terms. Furthermore, infinite results often crop up in calculations involving the Standard Model. An ingenious mathematical process called renormalization had to be developed to avoid these embarrassing divergences. Renormalization involves subtracting one infinite series from another and keeping only the terms that match the known result.

Many theoreticians criticize the Standard Model because of these mathematical difficulties, calling it inelegant. Perhaps physics' dissatisfaction is related to the philosophical presupposition that says the universe should not only be knowable, but our knowledge should be mathematically simple, tidy, and self-contained. Of course, these criticisms don't affect the remarkably precise agreement between prediction and experiment, nor do they hinder the Standard Model's explanation of many phenomena of the universe. The discomfort, however, keeps theorists searching for a better theory.

At a deeper scientific level, the Standard Model has a fundamental weakness, referred to as electroweak symmetry breaking. The photon, a massless boson, carries the electromagnetic part of the electroweak interaction. To preserve symmetry, the weak part of the electroweak interaction should also be carried by a massless boson. It isn't. The weak interaction's carriers are two W's and one Z, all of which have substantial masses—way more than most quarks. The symmetry is broken, and the Standard Model has no explanation for this result.

The most serious criticism of the Standard Model has to do with gravity and the origin of mass. The Standard Model ignores gravity entirely, and requires that particle masses, charges, and several other properties be measured experimentally in order to be substituted in the equations. Whatever theory replaces the Standard Model must address these criticisms and yet must correspond to the Standard Model in areas where its predictions match reality so well.

The Origin of Mass Problem, a.k.a. the Higgs Field Problem

On a purely mathematical basis, in 1964 Scottish physicist Peter Higgs and others postulated the existence of an all-pervasive field, later called the Higgs field. All particles that interact with the Higgs field acquire mass because of their interaction with that field. In other words, all mass comes about because of interaction.

This mass-conferring mechanism has been likened to soldiers marching through molasses. They become massive because molasses sticks to them as they march. Another analogy is that of a cocktail party where the guests are uniformly distributed around a room. When a very important person walks in, her nearest neighbors cluster around her, increasing her effective mass. The more important the person, the more clustering increases her mass.

According to the theory, different particles have different couplings with the Higgs field, granting large masses to the W's and Z and zero mass to the photon and gluon. If the Higgs mechanism truly confers mass on fundamental particles, it provides at least a partial answer to the unsolved problem about where mass comes from.

But how can we tell if the Higgs field is real or just a mathematical convenience? Here's how: A sufficiently large jolt, such as an extremely energetic particle collision, within the cosmic molasses known as the Higgs field can set the molasses quivering. This field vibration can be detected. There should be a Higgs particle that "carries" the Higgs field in the same sense that a photon "carries" the electromagnetic field.

In the simplest theory, only one Higgs particle "carries" the Higgs interaction. More complicated theories have multiple Higgs particles, but there's still one particle that is lightest. It is possible that this lightest Higgs particle lies within the range of current accelerators.

For several years, CERN—the European laboratory for particle physics in Geneva, Switzerland—searched for the Higgs particle in an accelerator known as LEP (the Large Electron Positron collider). A possible and tantalizing event at 115 GeV (see the table on pages 26–27 for particle masses) was recorded, but more data are needed to guarantee that the background wasn't interfering. In 2001, CERN shut down the accelerator to build a more energetic one using the same tunnel. The new accelerator, called the LHC (Large Hadron Collider), is scheduled to begin operation in 2005 and will have enough energy (8,000 GeV/beam) to be a much more effective probe. Since March 2001, the Fermi National Accelerator Laboratory in Batavia, Illinois, has been searching for the Higgs in its Tevatron (1,000 GeV/beam), but events that will reveal the

"YOU SAY THIS PARTICLE —THIS HIGGS BOSON—
IS MISSING..."

Higgs are so infrequent that it may take a long time to find enough data to be statistically significant. The Superconducting Super Collider (SSC), approved by President George Bush in 1987, had the Higgs search as its principal objective and would have had plenty of energy (20,000 GeV/beam) and beam strength for the task. It was canceled, however, by the U.S. Congress in 1993.

Anticipating the results of the Higgs search, if the Higgs particle is found, and its mass is within range of current accelerator capability, the Standard Model can be extended to include its effects. This won't solve the origin of mass problem or resolve all the Standard Model's difficulties completely, but it's a start.

If the Higgs particle is found and has a mass beyond the anticipated range, the Standard Model breaks down, because it forecasts events as being more than 100% probable. This situation will necessitate a substantial overhaul or replacement of the Standard Model.

If multiple Higgs particles are found, new theories beyond the Standard Model will be needed.

If *no* Higgs particle is found, this again necessitates replacement of the Standard Model. Such new theories are discussed in the next section.

Thus, finding the Higgs particle, or at least establishing a lower bound for its mass, is critical to beginning to understand the crazy-quilt pattern of particle masses. However, some theorists think that the Higgs field is only a Band-Aid and won't really solve the fundamental origin of mass problem. The Higgs has been proclaimed as the rug of ignorance, under which the fundamental problems of the Standard Model are swept.

The Standard Model's omission of gravity is another facet of the unsolved mass problem. A straightforward way to proceed would be to devise a quantum gravitational theory. The best gravitational theory is Einstein's general Theory of Relativity, so why not just apply quantum principles to general relativity? Because it's not that easy. General relativity is a classical theory about how the smooth, large-scale geometry of the universe is connected to mass. It works well at large distances, but no extensive experiments have been carried out below particle separations of 1 millimeter (mm). This means that gravity's strength is simply extrapolated to the microworld. On the other hand, the Standard Model quantizes fields into discrete particles and operates on a very small scale. So when theorists tried to quantize general relativity, the theory produced infinite results for quantities that are clearly finite.

Another difficulty lies in gravity's extraordinary weakness relative to the other forces. In order to be on a par with strong and electroweak interactions, gravity must be of similar strength. This is referred to as the hierarchy problem. A huge energy gap exists between the energies at which the Standard Model applies and the energy at which the weaker gravity finally becomes comparable in strength to strong and electroweak interactions. No one knows why this big energy gap exists.

New Physics Needed

As it stands, there is no experimental support for any theory beyond the Standard Model. However, many possible theories await testing. Here are a few:

GRAND UNIFICATION THEORIES (GUT) AND THEORIES OF EVERYTHING (TOE) These names are misleading because they promise more than they can deliver. In fact, they are umbrella terms that refer to an integration of the known interactions into a single, comprehensive theory. GUTs unite the electroweak and strong interactions. The more ambitious TOEs include not only strong and electroweak interactions but the gravitational interaction, as well. Even if such a theory is developed, it would hardly portend the end of science. A host of other problems confront science. This book is full of them.

M-THEORY Princeton physicist Edward Witten says "M stands for 'Magical,' 'Mystery,' or 'Membrane,' according to taste." Several earlier theories have been shown to be subsets of this overall theory—so-called strings, superstrings, and brane theories. Instead of treating quarks and leptons as point (one-dimensional) particles, this theory proposes that they have two dimensions (lines or strings) or even more dimensions (membranes, or "branes"). These related theories unify all forces, including gravity, and contain no embarrassing infinities that require renormalization like the Standard Model does. Since they all require more than four dimensions (10, 11, and 26 are the current major choices), the extra dimensions may range from being curled up too small to measure, within range of current measurement techniques, or even too big, approaching infinity. According to one theory, all the dimensions of the universe started out at similar size, but then separated and changed size as the expansion proceeded and the temperature cooled. One difficulty with choosing from among the many theories in this category is that our experience or intuition cannot be applied to dimensions beyond the four in which we live.

SUPERSYMMETRY (SUSY) If fermions and bosons are interchanged, the equations that describe fundamental interactions should still be valid. This theory predicts more massive superpartners for all existing particles. If superpartners exist, one or more might have a mass small enough to turn up in the search for the Higgs boson. The supersymmetric partners might also explain dark matter (see chapter 6). (Superpartners are named by applying an "s-" prefix for fermions; an electron's superpartner would be a selectron, a proton's would be a sproton, and so forth. An "-ino" suffix would be added to name bosons' superpartners; a photon's superpartner would be a photino, a W particle's superpartner would be a Wino, and so forth.)

TECHNICOLOR This theory considers quarks and leptons to be composites of smaller particles. Since this theory predicts new particles, this idea might be testable.

TWISTOR THEORY In this theory, both the Standard Model and general relativity are reformulated using complex number representations of space-time. (A complex number is defined as $a + ib$, where $i =$

the square root of -1 and a and b are real numbers.) The significance of complex numbers in the real world is unclear—they cannot be applied to count or measure any real entities.

As usual, to avoid being cast into the scrap heap of discarded theories, any scientific hypothesis must make predictions that are supported by experimental evidence. Some of these new theories are too abstract to make any predictions that can be tested; some are too difficult to allow calculations to be made; others involve quantities too far removed from our familiar surroundings for us to apply any constraints based on our experience or intuition. To provide experimental evidence for some of the extremely massive particles predicted would require an accelerator as big as the whole solar system.

Niels Bohr's correspondence principle, from the 1920s, said that quantum mechanics must agree with classical physics in situations where classical theory is known to be accurate. Applying that same logic to this case, any new theory must reduce to the Standard Model in situations where experimental evidence has supported the Standard Model. It could be a while before such a theory is found.

New Language Needed?

Although the descriptions of the Standard Model and possible replacement theories are purely conceptual, don't be misled. The language in which the Standard Model is expressed is mathematics, and that language itself may be inadequate. New mathematical concepts may be needed. To explain motion, Newton devised calculus, which works with smoothly varying functions and small numbers. We now recognize that the universe consists of noncontinuous functions and large numbers, but many of the equations are still framed in calculus terms. (More later in chapter 5 about weather prediction, which has similar difficulties.) Many of the theories that strive to replace the Standard Model involve mathematical concepts on a deeper level than calculus, such as groups, rings, ideals, and topological structures. Writing equations that describe the universe's functioning is one thing. Solving these equations in physically accurate and meaningful terms is something else.

Solving the Puzzle: How, Who, Where and When?

HOW In summary, we still don't know how the universe's fundamental building blocks got their mass, and we're not even sure we have identified all the building blocks. Nevertheless, there are both theoretical and experimental possibilities for deepening our understanding.

WHO On the theoretical side, many theorists are hard at work and may make a spectacular breakthrough at any time. To name just a few, watch for the ideas of Edward Witten, Frank Wilczek, Michio Kaku, M. J. Duff, Roger Penrose, Gordon Kane, and Lee Smolin.

WHERE AND WHEN Experimentally, the Higgs particle search is ongoing at Fermilab and will resume at CERN in 2005. Other facilities may be designed and built after that.

To keep track of these developments, follow the links in the "Resources for Digging Deeper" section. Future findings promise to be interesting, enlightening—and quite possibly surprising.

CHAPTER THREE

CHEMISTRY

By What Series of Chemical Reactions Did Atoms Form the First Living Things?

It is mere rubbish thinking, at present, of origin of life; one might as well think of origin of matter.

—Letter from Charles Darwin to J. D. Hooker, March 29, 1863

Chemistry is the study of the composition of substances and the transformations they undergo. The chemistry of both nonliving and living entities has been studied extensively, but the chemical transition from lifeless substances to that complex system of interacting molecules that display those behaviors we call life remains chemistry's biggest unsolved problem.

Primordial Soup

The right ingredients. In the right quantities. Mixed together at the right temperature. For the right amount of time. Depending on the ingredients, quantities, temperature, and time, this could be a recipe for oatmeal or a birthday cake. Or it could describe a primordial soup, stocked with certain organic molecules. When combined, these primordial molecules form larger replicating molecules made of proteins and nucleic acids. Creation of these larger replicating molecules, in effect, results in the creation of the genetic code, which is tantamount to the creation of life itself.

This chapter deals with the interface between chemical, or prebiological, evolution and biological evolution. It deals with the ingredients, quantities, temperature, time, and reaction sequences involved in a transition that occurred between 4.5 and 3.8 billion years ago and discusses the question of how a lifeless planet produced its first life-form.

Chemical Systems Evolve

However leptons and quarks managed to obtain their masses, the job got done, and the Big Bang proceeded. As the universe expanded and cooled, quarks combined to form protons and neutrons. Enough nuclear fusion took place to produce helium nuclei to the tune of 25% of the matter in the universe. The remaining matter was in the form of protons. As time progressed, giant gas clouds coalesced gravitationally, forming galaxies and stars. In the cores of these stars, atomic nuclei more massive than helium were produced. Upon completion of their life cycle, some of these stars exploded and blew many of their nuclei out into the interstellar material, where most of them managed to attract electrons and build the form of matter we've grown to know and love—atoms. More time passed, and some of the atoms found themselves in great clouds called nebulae, which coalesced because of gravity and formed more stars as well as many smaller bodies, including planet Earth.

Atoms shared electrons with other atoms and formed *molecules.* Questions about whether certain atoms combine or don't combine, the number of each of the atoms involved, the speed of the reaction, the amount of energy needed or produced, and various other considerations are studied by the discipline of *chemistry.* Chemical changes are represented in the form of equations:

Reactant atoms and molecules	combine to form ⟶	Product atoms and molecules

Although chemistry has solved many intricate mysteries about atomic and molecular combinations, so far one major puzzle has eluded its grasp: By what series of chemical reactions did the early Earth's atoms develop into the complex system of interacting molecules that display those behaviors we call life?

One atom, *carbon,* is key to the complexity of living systems. Carbon's electron structure requires it to share electrons (covalently bond) and form a total of four bonds, or shared pairs of electrons. They may be single, double, and even triple bonds. Furthermore, carbon atoms combine readily with other carbon atoms. This flexibility in structure formation leads to molecules that take many forms and shapes, ranging from very simple to extremely complex.

The study of carbon compounds is called *organic* chemistry because it was once thought that only living (organic) systems could produce such molecules. We have since learned that these compounds may be made artificially as well as by living systems. Carbon-based molecules may have started out as relatively simple ones, but carbon's bonding capabilities then enabled it to make progressively more and more complex molecules, eventually leading to that complex system we call life. This process can be represented in chemical equation form in which each arrow represents a series of chemical reactions:

Simple atoms → Simple molecules → Complex molecules → More complex molecules → Even more complex molecules → System of highly complex molecules → Complex system called life

Hypothesizing about Life's Origins

We've framed the problem of the origin of life as a chemical puzzle, but that's certainly not the only possibility. Historically, people have had a lot of other ideas about how life on Earth came about. Many of these are far removed from those of chemistry. We'll begin by examining some of the notions about the origin of life from a historical perspective. Then we'll look at what chemists have learned about the question so far. Finally, we'll explain why chemistry still regards the origin of life as an unsolved problem.

HYPOTHESIS 1: LIFE'S ORIGIN WAS SUPERNATURAL Before there was any organized study of chemistry or even a scientific method that demanded evidence, there was a widely held hypothesis in the western world about the origin of life. Life-forms were put on Earth by supernatural or divine forces. This belief is known as special creation.

Beliefs involving supernatural or divine forces raise a fundamental difficulty: mixing religion with science. Religious beliefs are based on faith, which is an individual, subjective decision. In contrast, science is built on objective evidence. Technically, the two endeavors are so disparate that their ideas cannot be compared legitimately. Yet we are human, and we do compare.

What makes the origin of life question especially difficult is that it happened so long ago that it is impossible to study any direct evidence. It's like a solitary golfer who shoots to a blind par 3 hole and, when he gets to the green, discovers the ball in the cup. A hole-in-one? Maybe. He can't go back in time to find out if the ball went in by itself—or if it was assisted by a devilish jokester.

On a more scientific note, recall the example from chapter 1 of the strong X-ray source detected near the star HDE 226868 by the satellite *Uhuru* in 1971. Since HDE 226868 is over 8,000 light-years away, it is inaccessible for direct measurement. How, then, do we know the X rays indicate the presence of a black hole and not possible signals from an extraterrestrial civilization? If these two explanations are regarded as competing hypotheses and no direct experimental evidence is available, Ockham's Razor must be applied. The black hole hypothesis is simpler. It requires only existing physical principles and so is preferred. The subsequent discovery of many similar X-ray sources helps support this choice.

Thus, science formally rejects the role of God in the origin of life not only because of lack of evidence but also because God violates Ockham's Razor by being nonphysical. Many scientists believe in God, but when they wear their scientific hats, they must abide by science's rules. By the way, if life is ever discovered elsewhere in the universe, there will undoubtedly be interesting repercussions for religious beliefs as well as science. (See Idea Folder 4, Extraterrestrial Life.)

HYPOTHESIS 2: COMPLEX LIFE-FORMS WERE GENERATED SPONTANEOUSLY Long ago, people observed tiny frogs hopping around decaying logs, rats in sewage and garbage, and maggots crawling on old meat. In about 1620, Jan Baptiste Helmont, an early Belgian chemist (and alchemist), gave his recipe for mice:

> for if you press a piece of underwear soiled with sweat together with some wheat in an open mouth jar, after about 21 days the odor changes and the ferment coming out of the underwear and penetrating through the husks of the wheat, changes the wheat into mice. But what is more remarkable is that mice of both sexes emerge [from the wheat] and these mice successfully reproduce with mice born naturally from parents. . . . But what is even more remarkable is that the mice which came out were not small mice . . . but fully grown.

The hypothesis that complex multicellular living things can arise directly from nonliving materials is often referred to as spontaneous generation, although some wags have referred to it as "logs to frogs". Before we derive too much mirth at the expense of 400-year-old ideas, we should pause for at least a moment to recognize that our own thoughts will likely look equally naive 400 years from now.

Once the Scientific Revolution occurred, and experimentation was installed as the final test of a hypothesis, Galileo's countryman and successor at the Medici court in Florence put spontaneous generation to the test. In 1668, Francesco Redi conducted an experiment in which he placed pieces of meat in various containers. Some were open to the air, some sealed completely, and others covered with gauze that had mesh with openings too small to allow anything to pass besides air. The inevitable flies buzzed around each container, but since maggots (fly larvae) appeared only in the open containers (into which flies could lay their eggs), it demonstrated that maggots came from flies and not from the meat. Furthermore, eggs were found deposited on the gauze. While you might imagine that Redi's demonstration would destroy the spontaneous generation theory completely, it did not. Many people still believed in it. Old ideas die hard. Even Redi continued to believe spontaneous generation could occur under some circumstances.

Soon after Redi's work came the invention of a powerful new experimental tool: the microscope. This was a mixed blessing. Not only was it extremely useful for biological observations, it also rekindled spontaneous generation beliefs because the "animalcules" that the microscope revealed appeared to arise spontaneously.

Around 1860, Louis Pasteur accepted a challenge that made a giant difference in the ongoing spontaneous generation controversy. A fellow member of the French Academy of Sciences, F. A. Pouchet, had published results of experiments in which he claimed to be able to spontaneously generate organisms at will. Pasteur pointed out some flaws in the experimental methods while Pouchet petitioned the French Academy to offer a prize for anyone who could prove or disprove spontaneous generation. Pasteur's friends advised him not to try for this prize, because it was designed to embarrass him. But Pasteur's earlier experiments with fermentation had prepared him so well that he was undaunted.

He undertook a series of experiments that culminated in putting sterilized broth in flasks with S-shape necks. Because of the shape of the flasks' necks, air could get in but microorganisms could not. The broth remained sterile, which showed there was no spontaneous generation of microorganisms. Pasteur said:

> Gentlemen, I could point to that liquid [in the flask of sterile culture medium on the table before him] and say to you, I have taken my drop of water from the immensity of creation, and I have taken it full of the elements appropriated to the development of inferior beings. And I wait, I watch, I question it, begging it to recommence for me the beautiful spectacle of the first creation. But it is dumb, dumb since these experiments were begun several years ago; it is dumb because I have kept it from the only thing man cannot produce, from the germs which float in the air, from Life, for Life is a germ and a germ is Life. Never will the doctrine of spontaneous generation recover from the mortal blow of this simple experiment.
>
> No, there is now no circumstance known in which it can be affirmed that microscopic beings came into the world without germs, without parents similar to themselves. Those who affirm it have been duped by illusions, by ill-conducted experiments, spoilt by errors that they either did not perceive or did not know how to avoid.

Pasteur won the prize, but his brilliant experiment still didn't completely kill the spontaneous generation theory, which resurfaced again and again. Like modern urban legends, it had a life of its own. From a scientific standpoint, Pasteur's experiments were almost too good. For, if every living thing came from prior living things, how then did the *first* living thing come about?

HYPOTHESIS 3: LIFE CAME FROM OUTER SPACE Anaxagoras, a Greek who lived from 500 to 428 B.C.E., philosophized about the "seeds of life," which, according to him, are present in all organisms. His philosophy has been interpreted as the beginnings of the idea of panspermia, the concept that life on the surface of planets comes from somewhere in space. In the 1870s, Scottish physicist William Thomson, Lord Kelvin, who found carbon in meteorites, was inspired to say:

> When two great masses come into collision in space, it is certain that a large part of each is melted, but it seems also quite certain that in many cases a large quantity of debris must be shot forth in all directions, much of which may have experienced no greater violence than individual pieces of rock experience in a landslip or in blasting by gunpowder. Should the time when this earth comes into collision with another body,

> comparable in dimensions to itself, be when it is still clothed as at present with vegetation, many great and small fragments carrying seeds of living plants and animals would undoubtedly be scattered through space. Hence, and because we all confidently believe that there are at present, and have been from time immemorial, many worlds of life besides our own, we must regard it as probable in the highest degree that there are countless seed-bearing meteoric stones moving about through space. If at the present instant no life existed upon Earth, one such stone falling upon it might, by what we blindly call natural causes, lead to its becoming covered with vegetation.

Working at the same time, German physicist Hermann von Helmholtz supported these notions, saying "It appears to me to be a fully correct scientific procedure, if all our attempts fail to cause the production of organisms from non-living matter, to raise the question whether life has ever arisen, whether it is not just as old as matter itself, and whether seeds have not been carried from one planet to another and have developed everywhere where they have fallen on fertile soil."

While these were interesting ideas and were expressed by leading scientists of the day, they weren't hypotheses from which predictions could be made and tested by experiment, so they went nowhere, at least in the scientific method sense.

In 1907, Sweden's Svante Arrhenius, Nobel Prize winner for his work on the chemistry of ions, wrote a popular book, *Worlds in the Making.* Arrhenius suggested that life began elsewhere, drifted up through some other planet's atmosphere, and migrated through space as spores driven by radiation pressure from whatever star was at the center of that planetary system. Viewed as a hypothesis, this idea predicted that spores would survive the Sun's ultraviolet radiation as they arrived at Earth. A number of experimenters tested spores under spacelike conditions and found they did not survive. As a result, Arrhenius's theory faded away—except as an inspiration for science fiction stories.

One of the biggest objections against panspermia is that it doesn't really answer the question of how life first came about. It just pushes the problem to some other, less accessible location. More recent variations on the panspermia theme will be explored later in this chapter.

HYPOTHESIS 4: LIFE ORIGINATED SPONTANEOUSLY RIGHT HERE ON EARTH In the 1920s, methane (CH_4) had just been discovered in the atmospheres of Jupiter and the other gas giant planets. Russian biochemist Aleksandr I. Oparin presumed methane had also been present on early Earth, along with ammonia (NH_3), hydrogen (H_2), and water (H_2O). These were presumably the kind of raw materials needed to start life going because they contained the key elements in life-forms: carbon, oxygen, hydrogen, and nitrogen. In 1924, Oparin published a pamphlet about the origin of life in which he stated:

> At first there were the simple solutions of organic substances, the behavior of which was governed by the properties of their component atoms and the arrangement of those atoms in the molecular structure. But gradually, as the result of growth and increased complexity of the molecules, new properties have come into being and a new colloidal-chemical order was imposed on the more simple organic chemical relations. These newer properties were determined by the spatial arrangement and mutual relationship of the molecules. . . . In this process biological orderliness already comes into prominence. Competition, speed of growth, struggle for existence and, finally, natural selection determined such a form of material organization which is characteristic of living things of the present time.

Oparin had discovered that a solution of proteins can collect together and form an aggregate. He called these aggregates *coacervates* and claimed they were capable of a kind of metabolism. Since there was a revolution going on in Russia in the 1920s, Oparin's works didn't become known in the West until the late 1930s.

In a 1929 essay, J. B. S. Haldane, a British biochemist, speculated about the origin of life on Earth. Citing recent experiments on the effect of ultraviolet radiation on chemical reactions, Haldane suggested that ultraviolet radiation acting on an early Earth atmosphere of carbon dioxide (CO_2), water vapor (H_2O), and ammonia (NH_3) could yield organic chemicals that would accumulate in the ocean and eventually reach the "consistency of hot, dilute soup." Subsequent chemical synthesis eventually yielded primitive organisms that fed on the organic nutrients around them. Haldane focused especially on the process of reproduction, and thought the primitive organisms would resemble simple viruses or viroids. Haldane's interests were quite wide-ranging, and his rationalist views well known. Near the end of his career, someone asked him what his long study of nature implied about its creator. Supposedly, Haldane paused to think—possibly about the 350,000 species of beetles, which account for over 50% of all insects—and then replied, "The Creator, if he exists, has an inordinate fondness for beetles."

Since Oparin and Haldane arrived at similar ideas independently, their hypotheses are often lumped together as the Oparin-Haldane theory. While their ideas are compatible, a key difference is that Oparin emphasized metabolism while Haldane focused on reproduction. This dichotomy continues to divide origin of life theories into two camps.

Once a hypothesis is proposed, it's only a matter of time before a testable prediction is made and a suitable experiment is carried out. In 1952, Stanley L. Miller (Nobel Prize–winner Harold Urey's graduate student at the University of Chicago) performed a seminal experiment designed to test the Oparin-Haldane theory. The presumed ingredients of Earth's primordial atmosphere—water, hydrogen, ammonia, and methane—were sterilized, put into an appropriate apparatus, and subjected to electrical discharges that simulated lightning. (See Figure 3.1.)

After the experiment ran for several days, Miller found simple organic molecules in the water. (See the table on page 48.) Among these molecules were amino acids, known to be building blocks of living organisms. (See Idea Folder 5, Amino Acids.) Of the many possible varieties of amino acids, only about one hundred occur in nature. Twenty are found in living organisms. *Four of these twenty were formed by random*

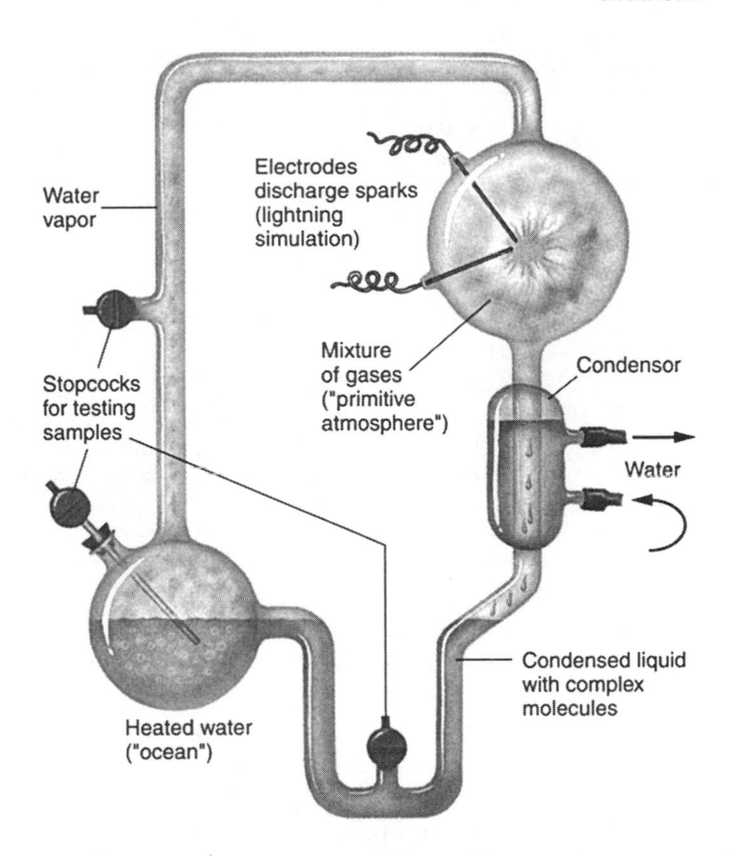

FIGURE 3.1. Miller Experiment Apparatus (from *Biology 6e,* by Raven & Johnson, © 2002 McGraw-Hill, used with permission of the McGraw-Hill Companies)

chemical reactions in Miller's apparatus. Large quantities of these simple yet distinctive organic molecules formed in only a few days.

These results provided impressive experimental support for the Oparin-Haldane theory. Of course, completely functional living organisms were *not* formed. Although the molecules that were present in Miller's apparatus were only simple components of molecules necessary for life, the fact that the molecules formed within days seemed to provide significant support for the theory.

The experimental support for the Oparin-Haldane theory of the origin of life was impressive but sketchy because the details of the biochemistry of life had not yet been worked out. Within a year, a profound development changed everything: At Cambridge, James Watson and Francis Crick determined the basic structure of the molecule that

Molecules Formed in Miller's Experiment

Molecule	Molecular Formula	Molecule	Molecular Formula
hydrogen cyanide	CHN	α-aminobutyric acid	$C_4H_9NO_2$
cyanogen	C_2N_2	α-aminoisobutyric acid	$C_4H_9NO_2$
cyanoacetylene	C_3HN	formic acid	CH_2O_2
formaldehyde	CH_2O	acetic acid	$C_2H_4O_2$
acetaldehyde	C_2H_4O	propionic acid	$C_3H_6O_2$
propionaldehyde	C_3H_6O	urea	CH_4N_2O
glycine	$C_2H_5NO_2$	aspartic acid	$C_4H_7NO_4$
sarcosine	$C_3H_7NO_2$	iminoaceticpropionic acid	$C_5H_9NO_4$
glycolic acid	$C_2H_4O_3$	succinic acid	$C_4H_6O_4$
alanine	$C_3H_7NO_2$	glutamic acid	$C_5H_9NO_4$
n-methyl-alanine	$C_4H_9NO_2$		
lactic acid	$C_3H_5O_3$		

holds the key to heredity: deoxyribonucleic acid, or *DNA*. As molecular biologists began to sort out the intricate dance between DNA, *RNA* (ribonucleic acid), proteins, and other molecules that enable living organisms to function, additional bits of information about the molecular interactions became known. As is often the case, the devil is in the details. The Oparin-Haldane theory of life's origins contained no detailed set of chemical reactions for the production of life because these molecules were not known at the time.

For the remainder of this chapter, we'll describe the current understanding of the molecular basis of an organism's functions. We'll project backward to what might have been the first, simplest form of life. Then we'll look at environmental conditions at the time of Earth's formation and examine how chemical reactions might have converted the simple molecules in that mixture into the molecular machinery that controls life. Next we'll explore some other complications that make the origin of life a major unsolved problem. Finally, we'll investigate a few avenues that might lead to its solution.

Life Now: Cell Structures

From the perspective of the present, life appears to be an extremely complicated phenomenon. With millions of species (350,000 of which are beetles), it would be difficult to expect that the simplest form of life would still linger nearby, available for our study. It didn't. With more than 4 billion years of mutations, reproduction, competition for food supplies, and environmental changes, it should come as no surprise that the first tentative form of life is long gone.

For that matter, what exactly is life? In 1947, the irrepressible British geneticist J. B. S. Haldane said, "I am not going to answer this question." After wrestling with borderline cases like viruses, viroids, and prions, biology moved beyond maintaining a hard and fast definition of life.

Living organisms have sometimes been characterized in terms of activities they need to perform:

Metabolize: Take in energy, use it, and release waste products
Grow and repair: Achieve appropriate size and fix imperfections
Respond to stimuli: Take action, based on external environmental occurrences
Reproduce: Make another organism like itself

Modern biology takes a simpler route: *Every living thing is cellular.* A particular organism may be a single cell or a complex mass of interacting specialized cells, but it is definitely cellular. Further, every cell has a membrane boundary to separate it from the rest of the world. Within this membrane is a complete set of instructions for the cell's operation and reproduction. These instructions are encoded in the molecule deoxyribonucleic acid, or DNA.

For quite a long time, it was thought that there are only two fundamentally different kinds of cells, *eukaryotes* and *prokaryotes.* (See Figure 3.2.) These differ in the location of their instructions (eukaryotes have a nucleus; prokaryotes don't) and their reproduction (eukaryotes reproduce by cell division, called mitosis; prokaryotes reproduce by simple cell fission). Recently it has been discovered that there is another category of cells, called *archaea.* Archaea resemble prokaryotes anatomically—both lack a nucleus—but archaea have some genes also found in eukaryotes and other genes that are completely unique.

Archaea's DNA is contained in a single circular molecule rather than the several stranded molecules that house the eukaryotes' DNA. Most archaea have a metabolism that doesn't require oxygen (anaerobic), and some (called extremophiles) thrive in environments way beyond the tolerance of any other organism. Hyperthermophiles, which live in water above the normal boiling point of 100 °C (212 °F), have been found in hot springs in Yellowstone National Park as well as near deep-sea thermal vents called smokers (more about these later). Others grow in cold, salty, or acidic environments, such as water pools under the Antarctic ice pack, saltwater lakes, and coal mining debris. Study of these phenomena has been an extremely exciting research area since the late 1970s.

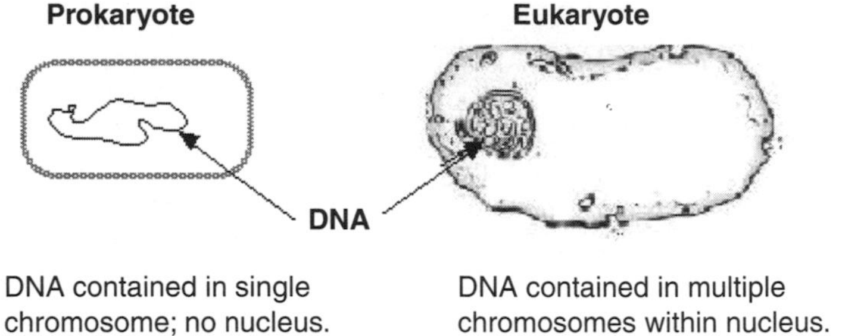

FIGURE 3.2. Prokaryote and Eukaryote Cells

Archaea are thought to be the oldest cells, predating both prokaryota and eukaryota. Thus, archaea may be closer in form to the earliest life than any other cell. Its lack of a nucleus and simpler DNA makes the archaea a good candidate for being a close relative of the first simple organism.

Cell Functions

Next, let's investigate the molecular functioning of cells. At a molecular level, a cell's genetic information is carried by the famous molecule DNA. (See Figure 3.3.) DNA is a relatively long double helical molecule, consisting of nucleotides joined in pairs. The junction between these nucleotides links pairs of nitrogenous bases, which join only in specific ways: adenine (A) joins only with thymine (T), and guanine (G) links only to cytosine (C). These are called Watson-Crick base pairs. The rest of the nucleotide is a sugar (deoxyribose) bonded to a phosphate, which forms the backbone of the spiral structure. (See Figure 3.4. Note that in molecule diagrams, when a ring corner shows no atoms at all, it is understood that this represents a carbon atom.)

The DNA molecule builds RNA molecules (messenger, transfer, and ribosome RNAs), which are single-stranded chains of nucleotides. RNA's nucleotides have the same structure as DNA, except uracil (U) replaces thymine (T). (See Figure 3.5.)

The double-stranded DNA is too big to fit through the openings in the nucleus's membrane in eukaryotes, whereas the single-stranded and shorter messenger RNA slips through nicely. Prokaryotes have no such problem because their DNA isn't enclosed in a nucleus. DNA replicates by splitting down the middle, then rebuilding the complementary halves of the molecule as a consequence of the attraction of Watson-Crick base pairs for each other. Both the splitting and the rebuilding functions require the assistance of enzymes (discussed shortly). RNA, transcribed

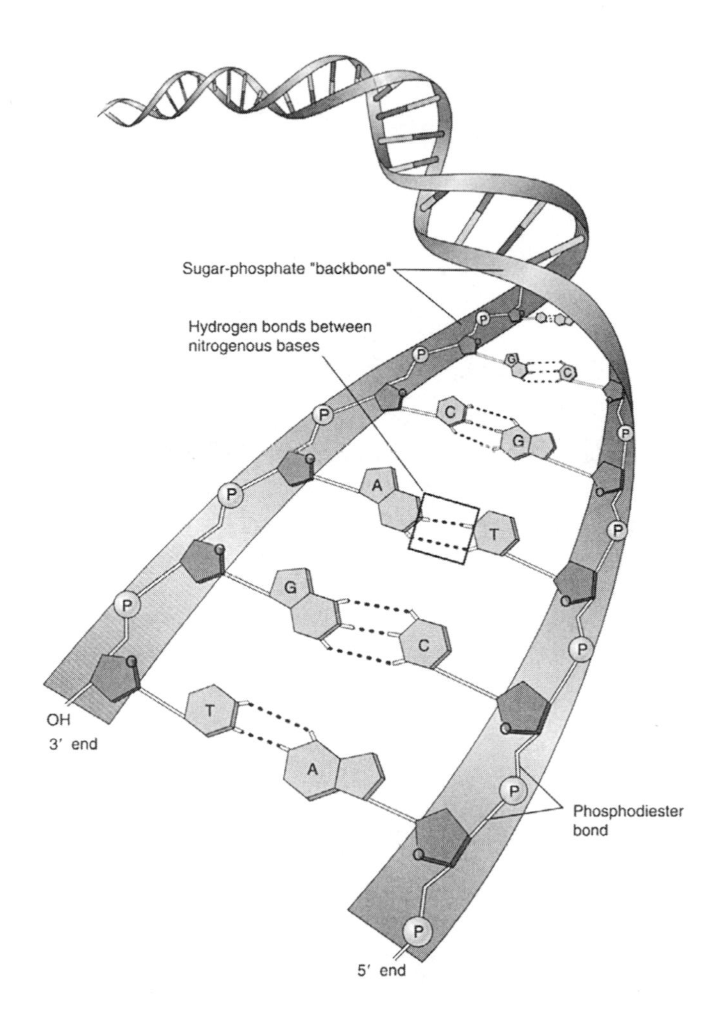

FIGURE 3.3. Structure of the DNA Molecule (from *Biology 6e,* by Raven & Johnson, © 2002 McGraw-Hill, used with permission of the McGraw-Hill Companies)

from DNA, then builds proteins, which consist of a long chain of amino acids. (See Figure 3.6.)

$$\text{DNA} \rightarrow \text{RNA} \rightarrow \text{Proteins}$$

Proteins carry out the cell's function by enabling certain kinds of chemical reactions to occur within the cell: reactions that build needed

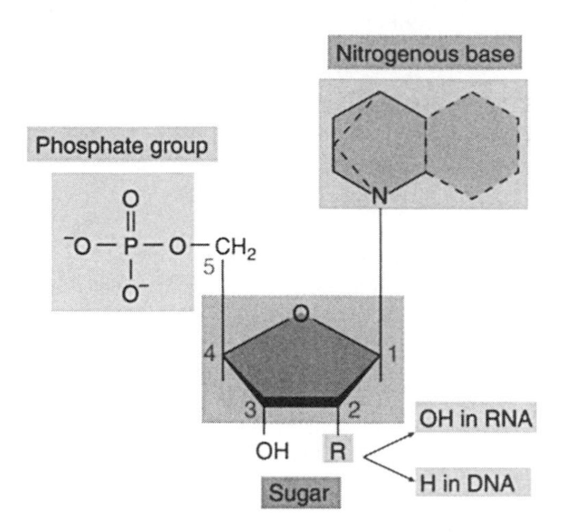

FIGURE 3.4. Structure of a Nucleotide (from *Biology 6e,* by Raven & Johnson, © 2002 McGraw-Hill, used with permission of the McGraw-Hill Companies)

parts, digest nutrients, store energy, and perform other housekeeping functions. (By the way, the details of the DNA → RNA → proteins operating system are not completely understood, especially the proteins and their folding, and constitute the biggest unsolved problem in biology; see chapter 4.)

To illustrate the function of protein enzymes, which facilitate only certain chemical reactions, consider the way the human body obtains energy: sugar and fats are oxidized. This same oxidation process happens in the external world. Did you ever see raw sugar burn or witness a grease fire? Both processes require extremely high temperatures, and yet the human body interior maintains a comfortable 98.6 °F (37 °C) while oxidation is occurring. The proteins built by RNA enable chemical reactions to occur at a much lower temperature, yet they are unaffected by the reaction and thus are not used up. Generally, such molecules are called catalysts.

In the case of biological molecules, catalytic functions are carried out by enzymes. Often enzymes bond complicated molecules temporarily. By slowing these molecules' motion, enzymes allow the molecules to link to other complicated molecules. This linkage is similar to the action of a key in a lock mechanism. As anyone returning home late at night can testify, it's much easier to get the key into the lock if the lock stays still. A catalyst also might mechanically fasten or unfasten the other molecules' bonds, then release them. The catalytic converter in automobiles is an example of a nonbiological catalyst. Finely divided particles of

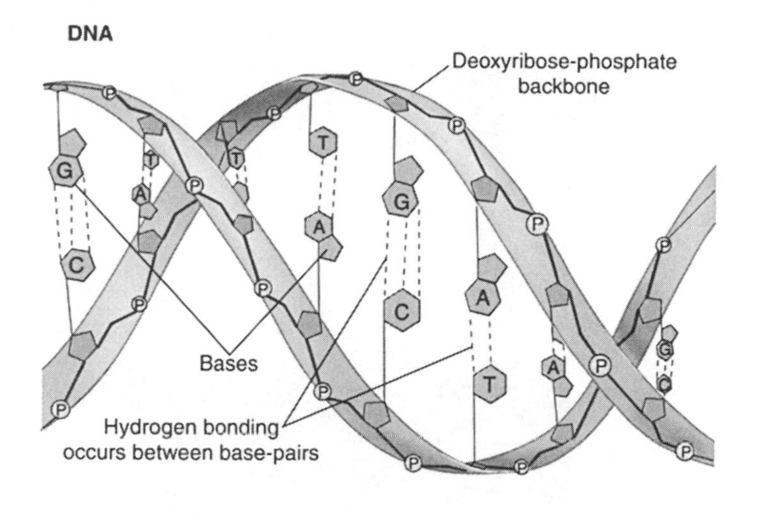

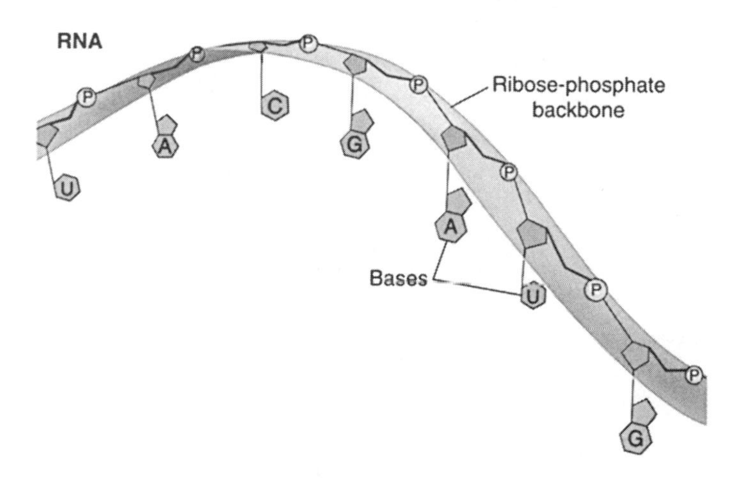

FIGURE 3.5. DNA and RNA Molecules (from *Biology 6e,* by Raven & Johnson, © 2002 McGraw-Hill, used with permission of the McGraw-Hill Companies)

platinum, palladium, or rhodium tear apart nitrogen oxides and release oxygen and nitrogen, combine carbon monoxide and oxygen to form carbon dioxide, or break apart unburned hydrocarbons ultimately to form carbon dioxide and water. In a sense, catalysts are like fight promoters who arrange for boxers to battle it out but don't take part in the contest themselves. (Think Don King.)

As you can see from the molecular diagrams, all these molecules are quite large and complex, yet they're built from simpler units. DNA

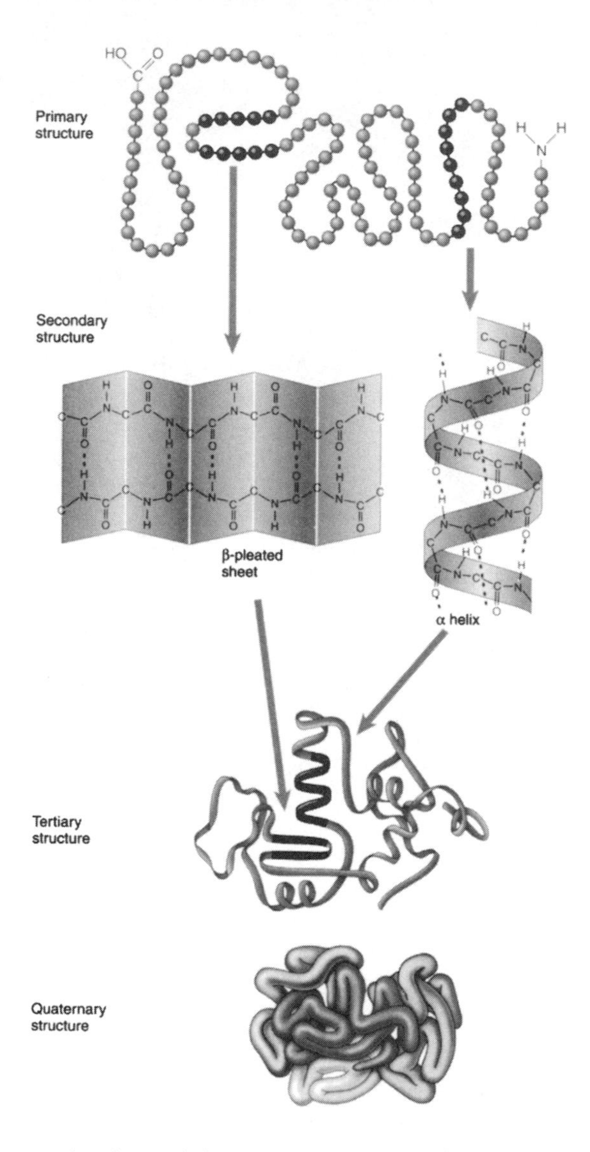

FIGURE 3.6. Protein Molecules and Their Structure (from *Biology 6e,* by Raven & Johnson, © 2002 McGraw-Hill, used with permission of the McGraw-Hill Companies)

and RNA are collections of *nucleotides,* each of which consists of a phosphate, a sugar (ribose or deoxyribose), and a nitrogenous base. Proteins are long, linked collections of amino acids. Each of these large linked collections is called a polymer. Just as a garden wall may be built in many shapes or sizes, depending on the size and shape of the stone blocks that make it up, a variety of different-shaped large molecules may be built from small ones fastened together. The single molecules are called monomers, and the process by which the small molecules are linked to form large ones is called polymerization.

One polymerization reaction is condensation, in which two monomers are connected while a water molecule between them is "condensed" out, thus linking the two monomers to form what's called a dimer (a "mer" of two parts). Three linked monomers are called a trimer, four a tetramer, and so on. Generically, when more than two monomers are connected, the product molecule is called a polymer (many parts). An example of a nonbiological condensation polymerization reaction is the curing of concrete. Silicate monomers form polymers, the excess water evaporates, and the gravel/sand mixture is enclosed in a silicate polymer framework. This results in a very strong bond.

In summary, DNA contains the plans for all proteins, including enzymes, and RNA builds the enzymes, some of which catalyze DNA replication. Enzymes cannot be built without the plans from DNA, and DNA cannot replicate without enzymes. Sounds suspiciously like the classic chicken and egg argument.

One way out of this dilemma was originally suggested by biochemist Leslie Orgel in the 1960s. RNA carried genetic information well enough, but if it also could act as an enzyme catalyst, it could fulfill the functions of both DNA and proteins. If this was the case, the original biomolecule wouldn't have to be either DNA or proteins, it could be RNA. Besides, RNA molecules are more easily synthesized than DNA, and DNA could possibly evolve from RNA.

Up through the 1970s, all enyzmatic functions had been observed to be handled by proteins. RNA had never been detected acting as an enzyme. Early in the 1980s, however, molecular biologists Thomas Cech and Sidney Altman, working independently, found that RNA *could* act as a catalyst. Close to 100 RNA enzymes, called *ribozymes,* have now been identified.

This discovery cast a whole new light on the origin of life problem. In a 1986 *Nature* article, Harvard molecular biologist Walter Gilbert coined the term *RNA World.* He wrote:

> The first stage of evolution proceeds, then, by RNA molecules performing the catalytic activities necessary to assemble themselves from a nucleotide soup. The RNA molecules evolve in self-replicating patterns, using recombination and mutation to explore new niches . . . they then develop an entire range of enzymic activities. At the next stage, RNA

> molecules began to synthesize proteins, first by developing RNA adaptor molecules that can bind activated amino acids and then by arranging them according to an RNA template using other RNA molecules such as the RNA core of the ribosome. This process would make the first proteins, which would simply be better enzymes than their RNA counterparts. . . . These protein enzymes are . . . built up of minielements of structure.

There are alternatives to the RNA World hypothesis, most notably the Protein-first hypothesis of biochemist Sidney Fox and the Clay World hypothesis of chemist A. G. Cairns-Smith. These theories have attracted less research interest, and their discussion will be postponed until we explore the RNA world in more depth.

PreSol

We'll begin our journey along the pathway toward life at the time when life's basic building blocks, atoms, were first taking shape. In order to see how Earth obtained its collection of atoms, especially carbon atoms, we've got to go back. *Way back.*

A long time ago, somewhere in our own Milky Way galaxy, there was a certain star, call it PreSol. PreSol had formed by gravitational condensation of a large cloud of hydrogen and helium from the interstellar medium. Like most stars, PreSol consisted of a core, where gravity pulls protons close enough for nuclear fusion to take place, and an atmosphere of gases that is heated by the core's energy output. During the first part of PreSol's lifetime, the core fused hydrogen nuclei (protons) to form helium nuclei (called alpha particles). The atmosphere glowed brightly with energy given off by this process.

After a while, the core's hydrogens were partially depleted. This lack of fuel caused the core to shrink and get hotter, which made the atmosphere expand and redden. Meanwhile, the shrunken core heated up sufficiently to start fusing three helium nuclei to form a carbon nucleus, in what is called the triple alpha process. Since PreSol had a large mass, its gravitational force was very strong, and the helium was consumed fairly rapidly. The core shrank again, the temperature increased again, and, as a result, new fusion reactions made elements heavier than carbon. Fusion happens in layers, so a large star's core resembles an onion, with different fusion reactions going on in each layer. The atmosphere expands and contracts somewhat, but it doesn't keep up with the core changes. The core works frantically, trying to halt the contraction due to gravity, and so more and more massive nuclei are fused. When the fusion process reaches the element iron, the game is almost over. Fusing iron isn't energetically profitable, and fusing heavier nuclei is even less so. When the inevitable core collapse occurred, it was spectacular. PreSol blew up, scattering some of the core and all the atmosphere into the interstellar medium. (What happened to the rest of the core? See chapter 6.)

Material consisting of 70% hydrogen, 28% helium, and 2% heavier elements, mostly carbon, shot out at high speed. As gravity slowed PreSol's ejected material, it enriched the interstellar medium with heavier nuclei.

While PreSol's story does explain the origin of the heavy nuclei that our solar system and Earth possess, there is another fact to consider. Large stars complete their life cycles pretty quickly on an astronomical scale—in millions to hundreds of millions of years. There could have been thousands of PreSols before our own solar system formed. So the cloud of gas and dust that condensed gravitationally to form us may have been enriched by nuclei built by many prior stars.

Our Sol

The early part of our Sun's story is like PreSol's, except our Sun has less mass than PreSol. Small stars have a longer lifetime because their smaller mass doesn't drive the fusion processes so rapidly. Therefore, our Sun will last longer and have a less catastrophic demise. Our main focus, however, is on planet Earth. Actually, Earth's formation was similar to a star's except for the fact that on Earth, there was so much less mass present in the coalescing particles that nuclear fusion never occurred. The coalescing particles collided and packed together, with denser materials sinking to the core and less dense ones floating toward the surface.

The gas and dust particles collided with each other, came together through a process called accretion, and ultimately formed a hot, early version of Earth. Accreted masses called planetesimals continued to rain on the surface of the newly formed Earth. It is possible that one large planetesimal hit Earth a glancing blow and gouged out material that formed the Moon and also set Earth rotating. Finally, the newly started Sun blew much of the debris farther out in the solar system. The inner planet zone became fairly clean, except for the occasional impacts from dirty ice balls flung around the system by close encounters with the massive outer planets. These ice balls are what we now call comets. Their trails are mostly water vapor and carbon dioxide caused by the Sun's rays forcing the ices to go from solid directly to the gaseous state—to sublime.

RNA Evolves

On newly formed planet Earth, the surface was rocky and hot. It was still being pelted by planetesimals and comet tails, which deposited carbon-laced water vapor and carbon dioxide. As the Earth cooled further, the water condensed and added to the comet-tail water to form the liquid oceans. The gaseous atmosphere probably consisted of gases released by volcanoes: water vapor (H_2O), carbon dioxide (CO_2), ammonia (NH_3), methane (CH_4), and a little left-over hydrogen (H_2) that hadn't been lost yet due to Earth's weak gravity. No significant oxygen (O_2) was present because what little amount there might have been would react chemically and thus not remain in the free state.

Subsequent chemical reactions could have begun on Earth in this context. In order to serve as the basis for life, such reactions should all be easily accomplished with readily available molecules under the conditions obtaining at the time. The reactions should be strong and vigorous. They should be robust. Starting with simple molecules and proceeding as far as RNA, we'll explore each reaction, looking for where and how it might occur and what positive or negative influences were exerted by the surroundings. In terms of timing, all reactions must have started near the end of the planetesimal rain period and finished before the oldest fossilized remains were formed. This gives a span of 100 to 500 million years, or about 10^{16} seconds.

Figure 3.7 shows the chemical reactions that must occur to produce RNA.

1. *Simple molecules react chemically to form amino acids, which are precursors to nitrogenous bases.* Stanley Miller's experiment (1953) produced a variety of organic molecules by random chemistry, some of which were amino acids, precursors to nitrogenous bases. Similar experiments were performed using different ingredients and ultraviolet light instead of electrical discharges. Yet, the results have been similar—all 20 of the amino acids found in living organisms have been built, in varying amounts. (See Idea Folder 5, Amino Acids.) This process could have begun in the atmosphere and then migrated to the ocean. Or it might have started deep in the ocean, near hydrothermal vents (called smokers), where the high temperatures would have provided energy and speeded up chemical reactions. Since there was no life yet, the molecules could accumulate in the ocean without being eaten by scavengers, as would be the case now.

2. *Simple molecules react chemically to form ribose.* Although this reaction has been observed, the complete chain of reactions that generates ribose as the main product has not yet been identified. In the reactions where ribose is a minor product, yields are mostly much too low to qualify as the robust reactions needed to build enough of life's early molecules. Perhaps researchers haven't uncovered the right reactions to produce the required ribose, or perhaps unique inorganic or organic catalysts were present. Rather than continue the Miller approach and cook simple molecules for longer and longer periods of time, researchers have jumped ahead and combined intermediate molecules to see how the rest of the operation might proceed.

3. *Simple molecules react chemically to form phosphate.* This is a straightforward inorganic reaction and is easily accomplished with phosphorus atoms supplied by weathering rocks.

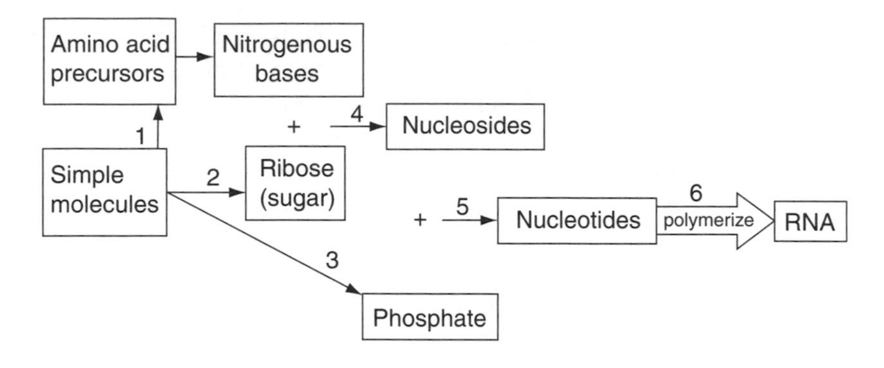

FIGURE 3.7. Chemical Reactions Leading to RNA.

4. *Nitrogenous bases and ribose react chemically to form nucleosides.* Researchers have achieved modest success in these reactions, but the synthesis is inefficient without the use of enzymes to catalyze the process.

5. *Nucleosides and phosphate react chemically to form nucleotides.* Some nucleotides are formed quite easily, while others are very difficult to produce. An additional stumbling block is that a large variety of nucleotides are formed. Some are normally found in organisms, but others are not and would interfere with RNA replication because they would not bond with naturally occurring nucleotides. Again, there may have been inorganic or organic catalysts helping this reaction. The catalysts may have originated on Earth or arrived from space on comet tails or meteorites. Their nature is currently unknown (more later). It's possible that nonenzymatic reactions occurred, but researchers haven't yet identified them.

6. *Nucleotide monomers polymerize to form nucleotide polymers—RNA.* Polymerization can be a difficult process in a water-rich environment. The soup might be too dilute; perhaps it needs to be more like stew or even pizza dough. Polymerization by condensation might have occurred in a shallow pond, on a sandy beach, or on a clay coastline. Large organic molecules could not survive much ultraviolet radiation, which suggests that shelter of some sort was required for polymerization to take place. It is conceivable that the water vapor molecules in the upper atmosphere were broken apart by the Sun's rays, in a process called photodissociation, yielding hydrogen and oxygen. The hydrogen escaped Earth's gravitational pull and the oxygen was transformed into Earth's first ozone (O_3) layer, which shielded the surface from ultraviolet rays. The oxygen was too high in the atmosphere to interfere with life-generating chemical reactions on Earth's surface and filtering the ultraviolet rays would keep organic molecules safe from being torn apart.

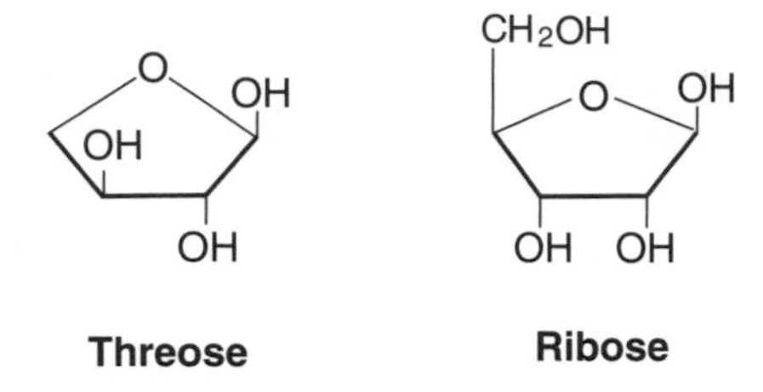

FIGURE 3.8. Four-Carbon Sugar (Threose) and Five-Carbon Sugar (Ribose) Molecules

(Atmospheric dynamics continued to evolve, helping to generate the unsolved problem about weather explored in chapter 5.)

Another possibility would be a self-replicating molecule that existed prior to RNA. This hypothetical precursor molecule would have to be easier to synthesize than RNA yet compatible with RNA's structure. Two possible candidates exist:

1. *TNA* (threose nucleic acid), which is built from four-carbon sugars (threose) rather than the five-carbon sugar (ribose) that forms the backbone of RNA. TNA could be synthesized more easily from the simple molecules in the prebiotic (before life began) world than RNA because it requires identical two-carbon fragments rather than the two- and the three-carbon fragments needed to form ribose. TNA polymers form double helices just like DNA and are compatible with DNA and RNA. (See Figure 3.8.)
2. *PNA* (peptide nucleic acid), where the backbone isn't formed by any sugar but consists of a polymer of the amino acid N-(2-aminoethyl) glycine. This molecule forms double helical structures, its components are easily synthesized by robust reactions of simple molecules, and it polymerizes easily.

Whether RNA had a self-replicating predecessor is unclear. While research on this question continues, let's explore the rest of the sequence.

RNA World

Once RNA is available, the route to the first cell becomes clearer. Five distinct stages of the RNA World development process remain.

1. *Replication stage*
   a. A strand of RNA constructs its complement (C-G, A-U) by the attraction of amino acid base pairs for one another. Every possible combination might form, but unstable ones wouldn't stick together as do the Watson-Crick base pairs (AU, GC), so those would dominate.
   b. The RNA complement separates from the original strand.
   c. The complement forms its own complement, which is identical to the original RNA.
   d. The molecular complements separate, resulting in a copy of the original RNA molecule and a complementary RNA molecule, both of which could construct more copies in similar fashion.

Duplication of all these steps under laboratory conditions has not yet been completely successful. It may be that catalysts aided these reactions. Inorganic catalysts in the form of charged clays may have assisted the process by attracting molecules and holding them in position for reaction. Another possibility is that self-catalyzing RNA molecules—ribozymes—may have performed the necessary replications. Still other organic catalysts may have been present, but haven't been discovered yet. Another difficulty has to do with the right- and left-handed nature of the helical RNA and DNA molecules, which will be discussed in the next section.

The possibility of Darwinian evolution at the molecular level exists in all stages of the RNA World development. Variation occurs during replication as a result of the random nature of the process. The various molecules produced would then compete for amino acids, and those molecules that were most efficient would corner the most amino acids and hence dominate. Note how this scenario is similar to the classic Darwin evolutionary process of variation, competition, selection, and amplification that operates at the level of organisms.

2. *RNA-directed protein synthesis.* RNAs that synthesized protein molecules must have enjoyed some Darwinian advantage, possibly through some indirect feedback loop that has not been identified yet.

3. *Partition into cellular groupings.* The formation of membranes from complex proteins or fatty lipid molecules must have occurred, separating groups of RNA from one another. This would foster competition between groups of RNA and protein molecules before they had reached the cellular stage. These cellular groupings are referred to as protocells.

4. *Linkage between proteins and RNA.* Theorizing that these first RNAs were divided into genes, each one of which synthesized one protein, it is estimated that they must have been 70 to 100 nucleotides long.

For comparison, today's human gene is several thousand nucleotides long. The early protein itself (actually a fragment, called a peptide) was probably 20 to 30 nucleotides in length. On theoretical grounds, the minimum number of genes has been estimated to be 256, so the first cellular RNA must have been roughly 20,000 nucleotides long.

5. *DNA information storage and formation of protein enzyme catalysts.* While RNA certainly stores genetic information quite capably, DNA's double helix is a more stable information storage structure than the single helix of RNA. Continuing with the idea of RNA building a variety of molecules in its role as both information storage and enzyme catalyst, it makes evolutionary sense that, once RNA built some DNA, the superior DNA genetic information storage would replace RNA information storage. Additionally, protein enzymes perform catalyst functions with more efficiency than RNA, so proteins would replace them in that function, as well. Thus, RNAs became the messenger, translator, and ribosome worker molecules because their other functions were subsumed by molecules that could do the job better. Darwin would be pleased. Once the protocell became capable of metabolism and reproduction, it was a full-fledged cell. Life had begun.

RNA World Alternatives

There are several other scenarios involving RNA, including Protein First and Clay World.

PROTEIN FIRST Sidney Fox showed in 1977 that certain mixtures of amino acids, when heated in the absence of water, polymerize to form proteinoids (short polypeptides with some catalytic properties). If the proteinoids are then put into water, they form membranes and resemble cells. Fox called these cell-like structures microspheres. Within these microspheres, proteins presumably catalyzed the formation of RNA and DNA.

CLAY WORLD In this hypothesis, radioactivity provides the energy for amino acids to polymerize in clay containing iron and zinc, which serve as inorganic catalysts for the formation of both proteins and RNA at the same time. This scheme was proposed by A. G. Cairns-Smith in 1982.

Much less research effort is being expended on these alternative hypotheses, but that might change if any compelling evidence for one of them is discovered.

Complications

As we have seen, the origin of life is a highly complex process. Many questions remain unanswered, such as the identity and relative proportions of the raw materials, the role of temperature, the amount of water available, whether catalysts were present or not, whether catalysts were inorganic or organic, their source, what chemical pathways actually occurred, and so on and so on. The fundamental difficulty remains that we can't go back in time to check the details.

Perhaps in exasperation, some people cut through the thicket and demand simpler answers, such as by regarding the whole process as statistical and making an estimate of the overall probability of occurrence. Many of these estimates have been offered, most colorfully illustrated by the comment of astronomer Fred Hoyle, who said the probability of life forming from simple molecules would be like saying a "tornado sweeping through a junkyard would assemble a Boeing 747 from the materials therein." Assembling a complex technological artifact from simple raw materials in a single event is more like "logs to frogs" than the multistep process outlined above. Besides, the overall process is decidedly *not* random. Catalysts speed the reactions, and the Darwinian system of variation, competition, reward, and amplification of successful molecules makes the chemistry far more efficient than a random process. It would be more like fitting parts together many, many times and retaining anything that began to resemble a 747. With that kind of feedback, you just might eventually produce that 747.

Another difficulty is right- and left-handed molecules. Carbon's ability to form four bonds allows it to form three-dimensional tetrahedral structures. So a single carbon atom, even though it bonds with the same atoms, can form two distinctly different molecules, called stereoisomers. (See Figure 3.9.) While these molecules are mirror images of each other, because of their three-dimensional nature, they are not interchangeable. Anyone who has tried to put a left glove on a right hand knows this.

This handedness of molecules is called chirality. Since the molecules are too small to see, their chirality is determined by passing polarized light through a solution of them and noting the direction in which the plane of the light's polarization is rotated. Molecules that rotate the light to the left are designated L-, and those that rotate the light to the right are designated D-. A more complex nomenclature system is used for more complex molecules. A mixture of both L- and D-forms of the same stereoisomer is called racemic. The fact that stereoisomers exist as mixtures would amount to little more than a mildly interesting academic point except for the fact that biological systems are very sensitive to chirality. For example, the L-form of a particular molecule of ketone called carvone smells like caraway seeds, while the D-form of the same molecule smells like spearmint.

Even more important, the molecules in living systems maintain their chirality. Proteins contain only L-amino acids, no D-amino acids, and DNA contains only D-sugars, no L-sugars. This fact could indicate that all prebiotic chemistry emanated from a single source. Recent experiments, however, show that peptides of the same handedness (homochiral) replicate more readily than mixtures of the two (heterochiral) and even suppress the minority chirality in mixtures. So it might be that L-amino acids and D-sugars happened to be in the majority and happened to suppress their opposite numbers as replication progressed.

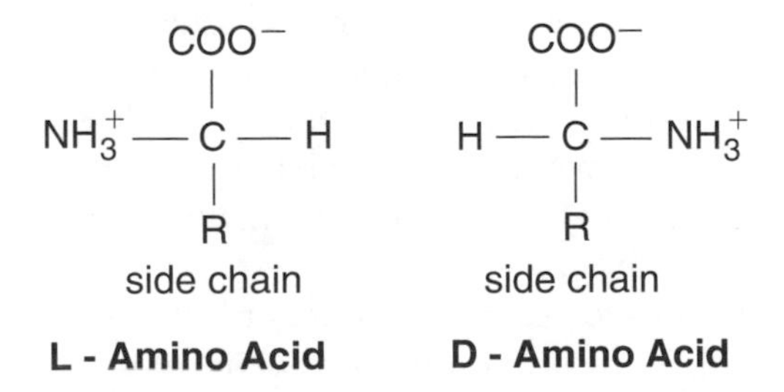

FIGURE 3.9. Chiral Stereoisomers

Another complication: Panspermia is back. In the 1960s, American astronomer Carl Sagan revisited Arrhenius' ideas and worked out the conditions under which small particles, such as spores, could travel through space. Rather than Earth, it turns out that satellites of the outer planets (e.g., Neptune's Triton, which has an atmosphere, or Jupiter's Europa, which has subsurface water) would be the best places in this solar system for any such spores to survive. This is not helpful for understanding the origin of life on Earth, but it does give a target for space exploration.

The next suggestion was made by British/American astronomer Thomas Gold, also in the 1960s. If some advanced civilization had explored this planet in the past and left remnants of their visit, this debris might be living and thus influence the course of development of life here. This "alien picnic theory" had no predictive power, but it has influenced thinking about our own exploration missions to other planets.

British astronomers Sir Fred Hoyle and N. Chandra Wickramasinghe started studying the spectra of cosmic dust in 1978. They became convinced that their highly complicated spectra matched that of freeze-dried bacteria. In their view, bacteria live on the dust particles in giant gas and dust clouds in space. When such a cloud condenses to form a solar system, the dust grains become the nuclei of comets and rain bacteria onto developing planets. The location and chemistry of the development of bacteria in the first place isn't clear, but at least this theory does allow more time for the first cell's development than the few hundred million years for prebiotic chemistry in the Oparin-Haldane hypothesis.

Other researchers have found support for some aspects of the Hoyle-Wickramasinghe theory. Over 130 different molecules have been identified by the presence of absorption lines in the spectra of stars shining through dust clouds. The molecules in the dust clouds include sugars, vinyl alcohol, and other molecules of biological significance. The mechanism by which these complex molecules form in clouds of extremely low density is not at all clear. If a dust grain within a cloud acts as a catalyst, holding onto simpler molecules until they react to form larger ones, then how does the larger molecule escape? Collisions with other particles violent enough to shake the large molecule loose would be energetic enough to break the catalyst-molecule bond. This puzzle surely needs more work.

Meteorites also have yielded significant amounts of organic molecules. For example, 70 different amino acids have been found in meteorites. Eight of them are from the set of 20 biologically significant ones. A meteorite found in Murchison, Victoria, Australia, in 1969 yielded many complex organic molecules. Its amino acids were predominantly of the L-type found in biological systems here on Earth.

There is a significant test of comet and interplanetary dust in progress. In 1999, NASA launched the *Stardust* spacecraft, which will take samples from the tail of Comet Wild 2 and of interplanetary dust and return them to Earth in 2006 for analysis. A fascinating preliminary result is that the *Stardust* spacecraft has already encountered particles with a molecular mass of 2000. While these particles can't be specifically identified until 2006, they are undoubtedly carbon-based and are almost 10 times as big as any molecules noted so far.

Could an extraterrestrial enzyme have catalyzed some key reaction in the prebiotic soup? We'll have to wait and see what kind of stardust *Stardust* finds.

Solving the Puzzle: How, Who, and Why?

HOW Let's put the two major testable hypotheses about life's origins into a scientific method context.

Hypothesis 1
Hoyle-Wickramasinghe's Panspermia

> *Prediction:* If bacteria live on comet nuclei, there should be life or, at the very least, complex organic molecules elsewhere.
> *Experiment:* Mars and outer planet satellite exploration missions or possibly the *Stardust* spacecraft will tell the tale. If no life is found, the hypothesis must be modified or discarded. If life is found, then . . . One wild card: If the California-based Search for Extra Terrestrial Intelligence (SETI) project gets a signal that seemingly originated from intelligent life-forms, the origin of these life-forms will be of great interest and significance. (See Idea Folder 4, Extraterrestrial Life.)

Hypothesis 2
Oparin-Haldane's molecular spontaneous generation. As our discussion showed, this hypothesis is still incomplete. Many details still need to be researched.

> *Prediction:* When details of the hypothesis are worked out, some series of robust reactions should be specified, all of which could be duplicated in the lab.

"I THINK YOU SHOULD BE MORE EXPLICIT HERE IN STEP TWO."

Experiment: Scientists await the prediction so they can design suitable tests.

WHO Who in particular might be involved in completing the hypothesis and carrying out the difficult lab experiments? Here's a partial list of candidates: Sidney Altman, David Bartel, Ronald R. Breaker, Andre Brock, A. Graham Cairns-Smith, Thomas Cech, Christian de Duvé, Manfred Eigen, Andrew Ellington, Albert Eschenmoser, James P. Ferris, Iris Fry, Walter Gilbert, Harold Horowitz, Wendy Johnson, Stuart Kauffman, Noam Lahav, B. E. H. Maden, Peter E. Nielsen, Harold Noller, Leslie Orgel, Norman Pace, Kourosh Salehi-Ashtiani, Eors Szathmary, P. J. Unrau, Charles Wilson, and Art Zaug. Or it could be someone in an obscure place like the Swiss patent office, someone with the clarity of vision to understand the big picture as well as the small details needed to bring it all together.

WHY Why do scientists work on such large and forbidding problems like the origin of life? Many are motivated by curiosity, but there's an especially interesting inducement for this project. The Origin-of-life Foundation, Inc., will award the Origin-of-Life Prize to anyone who proposes "a highly plausible mechanism for the spontaneous rise of genetic instructions in nature sufficient to give rise to life." The award is $1.35 million—a decent motivational carrot. For details, see the Web site: *www.us.net/life.*

In 1862, Louis Pasteur undertook a fundamental challenge against the advice of his friends. He solved the puzzle handily and won the French Academy of Science prize for his efforts. What we need is a twenty-first-century Pasteur.

CHAPTER FOUR

BIOLOGY

What Is the Complete Structure and Function of the Proteome?

What is life?
It is the flash of a firefly in the night.
It is the breath of a buffalo in the wintertime.
It is the little shadow that runs across grass
And loses itself in the sunset.

—Dying words of Crowfoot,
Blackfoot orator and warrior, 1890

However Earth's transition from a lifeless planet to a life-bearing one may have occurred initially, it paved the way for the evolution of a planet teeming with a wide variety of life-forms. Biology is the study of those life-forms and the processes by which they live. Until recently, the biggest unsolved problem in biology was the wording of the text of the molecular blueprint, or genome, of individual life-forms. Now that the human genome and the genomes of other life-forms have been mapped and sequenced, the unsolved problem has shifted into a new stage: How do the protein molecules built from the directions provided by these genomes contribute to the structure and functioning of organisms? How do these protein molecules enable the incredibly complicated molecular interaction called life?

E. coli

Hurry up and eat. Hurry up and grow. Hurry up and reproduce. Hurry up and respond. For cells, hurrying is a way of life.

Somehow, molecules perform all these vital cell functions. According to the central doctrine of molecular biology, DNA's message is transcribed in the form of RNA, which then translates the message into proteins, long chainlike polymers with many different side groups strung out along a repeating backbone. These proteins, in turn, accomplish the task of keeping the cell's functioning on track.

Life's operating system wallops any Windows version. Life's tiny molecular package achieves its goals reliably, in a variety of environments, and with few crashes. Although biology has made great strides in understanding life-forms, the details of life's operating system are so intricate that they constitute the biggest unsolved problem in biology.

"I COULDN'T HELP NOTICING HOW WELL YOUR MOLECULES INTERACT."

To get a sense of the nature of this problem, let's analyze some of the complexities of the molecular dance that occurs when a relatively simple organism performs one of life's essential activities—the metabolic breakdown of a sugar molecule. The process we're going to look at was first analyzed in the 1960s by French scientists Jacques Monod, François Jacob, and André Lwoff. We begin by focusing on the tiny bacterium that lives (usually quite peaceably) in the large intestine (colon) of many animals, including humans. Its name is *Escherichia coli,* usually abbreviated *E. coli.* As one of the favorite organisms of biological researchers (for reasons we'll discuss later), *E. coli* has been studied extensively. One particular strain, K-12, is quite harmless and often used in laboratory work. Its complete DNA (its genome) has been mapped and found to contain 4,639,221 base pairs. From K-12's DNA, 89 RNAs are transcribed, which in turn build 4,288 different proteins. When this sturdy organism's molecular mechanism operates on a simple (single) sugar, glucose, and a few inorganic ions, it is able to synthesize every organic molecule required to metabolize, grow, respond, and reproduce. Because of its adaptability, this tidy little life-form thrives in glucose-rich culture media in biological labs around the world.

E. coli Operons

The molecular versatility of *E. coli* is closely tied to the presence of *operons*—genetic units that are situated on the DNA molecule, the chromosome, and consist of a cluster of genes with related functions. One particular operon is called a *lac* operon because of its key role in the metabolism of the sugar lactose. The *lac* operon contains three genes that are responsible for producing three proteins that import lactose into the cell and break it down into glucose and another sugar, galactose.

Let's see how the *lac* operon participates in the metabolic process when lactose is added to the normally glucose-rich culture media. Lactose, literally "milk sugar", is more complex than glucose. It consists of glucose and galactose, linked to form a single molecule, called a disaccharide. (See Figure 4.1.) After lactose is added to *E. coli*'s environment, everything continues pretty much as before. *E. coli* metabolizes the glucose while the lactose just sits there. But if glucose isn't replenished, it is eventually used up.

E. coli's response to this development is very interesting. For a while, things get quiet. *E. coli* doesn't metabolize the lactose, other metabolic activities decline, and cell division ceases. Looks like trouble for *E. coli.* But before long lactose starts being metabolized, and *E. coli* is back in business. Analysis of the chemistry of the cell reveals the presence of three

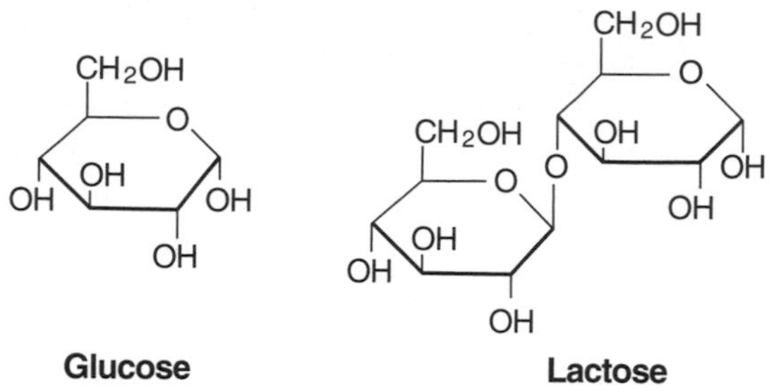

FIGURE 4.1. Glucose and Lactose Molecules

new proteins that weren't there before the glucose ran out. These proteins consist of a permease, which escorts lactose molecules across the cell boundary, allowing them to come in and be digested; beta-galactosidase, which breaks the lactose into glucose and galactose; and a transacetylase, whose function is not clearly understood.

Operon DNA → RNA → Proteins

You might imagine that the presence of lactose in the cell would simply act as a trigger, setting off the transcription of an RNA that produces these three protein enzymes. The actual process, however, is somewhat more complex. The signal to produce different protein enzymes involves both the presence of lactose and the absence of glucose. Let's view this process from the molecular level to see how it works.

DNA is sometimes thought of as a lonely, isolated molecule, secure because of its sturdy structure, protecting the cell's vital information. That's hardly the case. In fact, DNA is constantly being probed, poked, prodded, kinked, and unzipped by various protein enzymes. This molecule's activity would give the information superhighway a run for its money.

The basis for all the action is DNA's shape and electric charge distribution. The double helix has grooves, referred to as major and minor, and each of the nucleotide bases has unique electrical charge configurations. (Idea Folder 6, Building a DNA Model, provides information on how to build a portion of DNA from construction toys.) Some proteins possess exactly the right size and shape to "fit" into these grooves. Because of the electrical charge distributions of both the proteins and DNA, they may stick together fairly tightly. The attraction, however, is nowhere near as strong as the covalent bonds within each molecule. This fitting of one molecule into another is referred to as binding. Depending on their shape and charge distribution, proteins are bound at particular sites along the DNA. As the molecules jostle around in response to normal thermal motions, proteins are constantly bound and released.

The fitting together of complicated molecular shapes is often thought of in terms of a lock-and-key analogy. Only a few shapes fit together well enough to bind. Proteins also can bind to other proteins, which makes a new unit called a complex. Usually a complex has a different shape and charge distribution from the original molecule. This change is called conformational change, and plays a major role in the protein building process because it can cause "keys" to fit different "locks" or not fit the ones they fit before.

RNA is built with the help of a protein enzyme (a polymerase) that attaches to a binding site on DNA, opens the double helix right down the middle like a zipper, and transcribes the order of the DNA's base pairs onto an RNA molecule. The RNA then leaves the DNA and translates the nucleotide base pair order by building a protein at a piece of molecular machinery called a ribosome. Each group of three nucleotide bases, called a codon (see Idea Folder 7, Codons), specifies a particular amino acid to add to the protein. RNA polymerase binds to DNA only at sites where it fits. This fit is influenced not only by the polymerase molecule's shape but also by the availability of the DNA binding site, which, in turn, depends on the DNA's bends and kinks.

Three more molecules are needed to complete the description of the lactose metabolic process. The first is a catabolite activator protein, called CAP for short. In its normal state, CAP's structure doesn't allow it to bind to the DNA structure. CAP does contain a binding site for another molecule, cyclic adenosine monophosphate, popularly called cAMP (each letter is pronounced). The cAMP molecule is produced in an environment in which glucose is absent. If cAMP is bound to CAP, CAP undergoes a conformational change that allows it to bind to DNA. In turn, binding the CAP/cAMP complex to *E. coli*'s DNA causes DNA to bend, as Figure 4.2 shows.

The final step needs another protein, which functions as a repressor. In this specific case, it is called a *lac* repressor. This molecule normally fits into DNA's groove at a location where it interferes with the RNA polymerase that transcribes DNA's information into proteins that digest lactose. The *lac* repressor also has a binding site for lactose.

If lactose is not bound to the *lac* repressor, the repressor fits snugly in DNA's groove at exactly the right location to block the RNA polymerase from its transcription mission. If lactose is bound to the *lac* repressor, the *lac* repressor's conformation changes to the point that it no

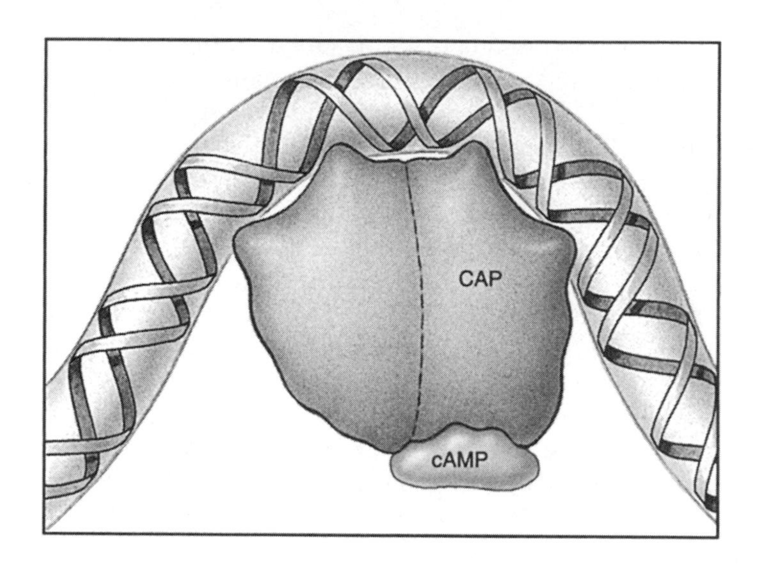

FIGURE 4.2. cAMP Bound to CAP, Bound to DNA (from *Biology 6e,* by Raven & Johnson, © 2002 McGraw-Hill, used with permission of the McGraw-Hill Companies)

longer fits on *E. coli*'s DNA, so the RNA polymerase is not blocked from its transcription function. Let's see how these molecules work together to produce the observed behavior of *E. coli*.

In the original environment, glucose was present and lactose absent. The presence of glucose meant that cAMP was not produced, so the CAP/cAMP complex wasn't formed, the DNA didn't bend, and the RNA polymerase didn't transcribe proteins for metabolizing lactose. Besides, the repressor was in place on the DNA, preventing RNA polymerase from being connected to DNA at that point of attachment. As a result, transcription was doubly prevented. (See Figure 4.3a.)

In the mixed glucose/lactose environment, the presence of glucose prevented the formation of CAP/cAMP, so the DNA didn't bend and the RNA polymerase didn't transcribe. Even though the presence of lactose took the repressor out of DNA's groove, RNA polymerase couldn't bind fully. This left the DNA without any attachments in the *lac* operon region.

If both glucose and lactose are absent, CAP/cAMP would be present, the DNA would be bent and ready to receive RNA, but the repressor also would be in place. From *E. coli*'s perspective, with no food available, this would be a hungry time. But notice how it is ready for action. If glucose shows up, it won't waste any energy making protein enzymes. It just starts digesting the glucose. If lactose suddenly appears, the DNA is kinked and ready to build the appropriate RNA as

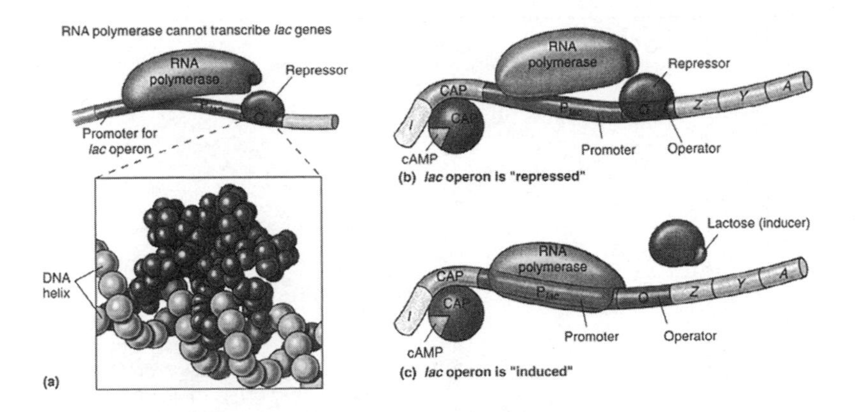

FIGURE 4.3. The *lac* Repressor (from *Biology 6e,* by Raven & Johnson, © 2002 McGraw-Hill, used with permission of the McGraw-Hill Companies)

soon as lactose binds to the repressor, which then shakes loose from the DNA. (See Figure 4.3b.)

When glucose is absent and lactose is present, everything lines up. The lack of glucose allows the CAP/cAMP complex to form. This complex is bound to DNA and causes the bend, which allows the RNA polymerase to find its binding site. The presence of lactose binds to the *lac* repressor and frees the repressor from the DNA, so the entire RNA polymerase can then bind to the DNA and build the three proteins that metabolize lactose. (See Figure 4.3c.)

By analogy, this situation is similar to a door equipped with both a doorknob and a dead-bolt lock. The doorknob acts like an activator, and the dead-bolt functions as the repressor. The following table compares the action of the doorknob/dead-bolt door system to the activator/repressor mechanism.

Doorknob Condition (activator)	**Dead-bolt Lock Position (repressor)**	**Will the Door Open? (Will lactose-digesting proteins form?)**
Not turned (high glucose)	Locked (low lactose)	no
Not turned (high glucose)	Not locked (high lactose)	no
Turned (low glucose)	Locked (low lactose)	no
Turned (low glucose)	Not locked (high lactose)	yes

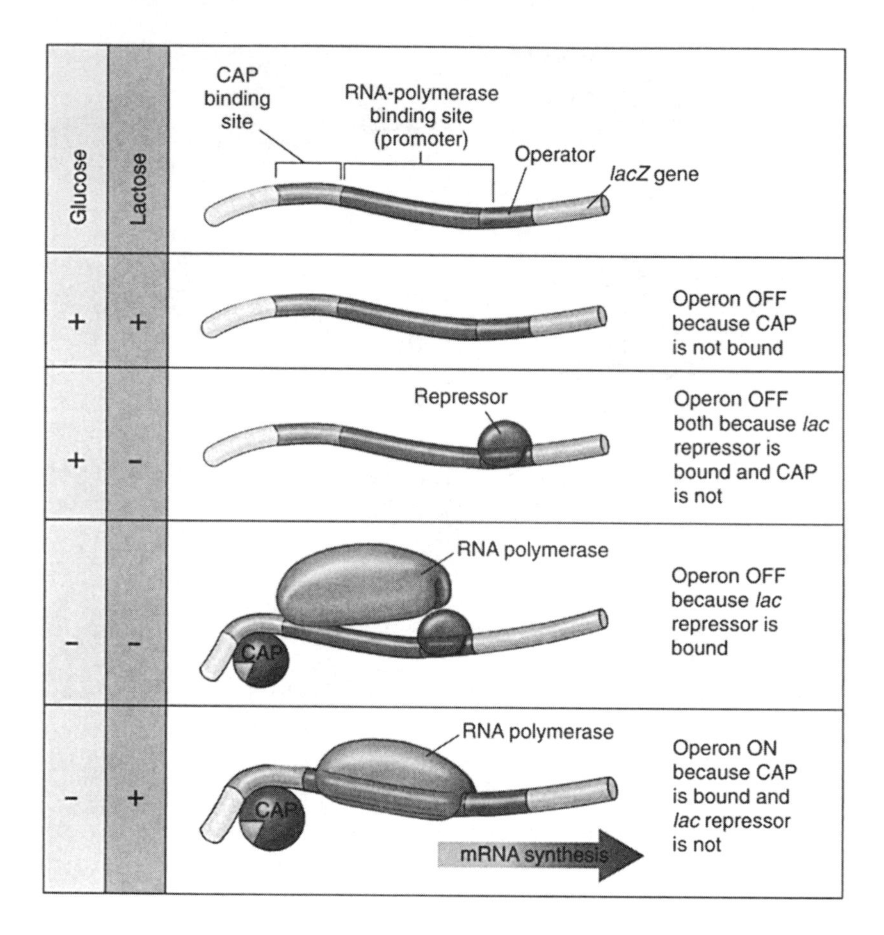

FIGURE 4.4. Summary of the *lac* Operon Molecular Action (from *Biology 6e,* by Raven & Johnson, © 2002 McGraw-Hill, used with permission of the McGraw-Hill Companies)

This intricate control system is like an old Rube Goldberg apparatus, in which a whole chain of complicated events is required to accomplish some simple goal (see Figure 4.4). Although it would be expensive in energy terms, why doesn't *E. coli* just make all the necessary enzymes continually so it could digest whichever sugar happened to come along? Perhaps just such a bacterium did exist at an earlier time. But when mutations produced an *E. coli* with this *lac* operon system, this *E. coli*'s efficiency was so much greater than its earlier relative that it crowded its cousin out and took over. That's classic natural selection at work.

Protein enzymes are built practically as soon as their RNA is transcribed, while the RNA is still attached to the long, circular, single-molecule DNA. Since *E. coli* is a prokaryotic cell, there's no nucleus or nuclear membrane to slow things down, so the digestion of lactose begins very quickly. *E. coli* thrives happily on lactose as well as glucose.

E. coli versus Other Organisms

E. coli is one of the best understood of all living creatures; scientists have identified almost two-thirds of the functions of its genes. The *lac* operon mechanism constitutes only a small part of *E. coli*'s molecular functioning. You may wonder why so much is known about this tiny bacterium, so small that 50 of them could fit across the diameter of a human hair. The answer is that it's much easier to conduct biological research using nonhuman subjects and doing so raises fewer ethical concerns. Simple organisms are easier and quicker than humans to breed, feed, and conduct experiments on. *E. coli* is certainly compact enough to qualify as a convenient research subject, and its reproduction rate is outstanding: It divides once every 20 minutes. Given enough water, glucose (or lactose), and enough room, one *E. coli* would become over 1 billion cells within 10 hours. Although other strains of *E. coli* are a health hazard, the K-12 strain is not, so extensive handling precautions are unnecessary. For over 70 years, *E. coli* has been a workhorse of biochemistry, genetics, and developmental biology. The similarity of its molecular processing to that of other organisms has even led to its use as a factory for the manufacture of insulin used by diabetic humans. Jacques Monod, eminent French scientist and 1965 Nobel laureate, once remarked, "What's true for *E. coli* is true for an elephant."

Other prokaryotic organisms have been studied, such as *Mycoplasma genitalium,* the smallest free-living organism, which has 580,000 base pairs and 517 genes in its DNA, and *Haemophilus influenzae,* which has 1,830,137 base pairs and 1,743 genes. But the relative simplicity of prokaryotic DNA in terms of its size and circular shape limits its applicability to more complex organisms.

From Prokaryotes to Eukaryotes

Prokaryotes live on the edge. These nimble little life-forms must respond quickly. When food is available, they must metabolize it and grow before the food runs out. A control system such as the *lac* operon is well suited to a rapid operation in which enzyme levels are adjusted up or down reversibly, responding to rapid changes in environmental conditions.

The situation with eukaryotes is very different. Most multicellular organisms have developed in such a way that their internal cells are insulated from transient changes in the surroundings. A constant internal

environment—homeostasis—is desirable for stable functioning of multicellular organisms. As a result, in eukaryotes gene control mechanisms are more slanted toward regulation of the organism as a whole. For example, some genes are activated only once and produce effects that are not capable of being reversed, as opposed to the flexible *lac* operon mechanism, which is fully reversible. In many animals, unspecialized cells, called stem cells develop extremely early in the embryonic state. These cells turn into specialized cells, such as brain cells or fingernail cells, by following a fixed genetic pattern, which may even lead to eventual cell death. This specialization of cells translates into more DNA, more RNA, and more protein enzymes so that eukaryotes can accommodate subtle interactions among these molecules in their metabolism.

Model Organisms

A eukaryotic organism that has been studied extensively is *Saccharomyces cerevisae* (*S. cerevisae*), commonly known as brewer's yeast. It is perhaps the best-understood eukaryotic organism at the molecular and cellular levels. Although *S. cerevisae* is only a single-celled fungus, many of its core cellular processes are similar to those in mammals. In fact, research on yeast has identified many molecules and chemical pathways that are involved in processes that go awry in cancer. *S. cerevisae* is more sophisticated than bacteria and has 6,000 genes within a DNA containing about 12 million nucleotide base pairs. Both *E. coli* and *S. cerevisae* are considered to be *model organisms,* which means that they:

1. Develop rapidly, with a short life cycle
2. Have a small adult size
3. Are readily available
4. Are easy to manage
5. Carry out biological functions in a manner similar to more complex organisms, such as humans

Other model organisms also have been studied extensively. *Caenorhabditis elegans* is a transparent roundworm that grows to about 1 millimeter in length, about half the size of this mark: ~. *C. elegans* achieves full growth in about three days, lives in the soil in many locations around the world, and feeds on microbes such as those found in rotting vegetation. This little worm is a multicellular (959 cells) eukaryote with 19,099 genes in its 97 million base pair DNA. It starts from a single cell and has a complex development process that forms a nervous system and a "brain." *C. elegans* exhibits behavior, is capable of learning, produces eggs and sperm, ages, loses vigor, and dies. Sydney Brenner, a molecular biologist from the United Kingdom, says that *C. elegans* is well named because it is "photogenic," as you can see in Figure 4.5. Brenner, John Sulston, and

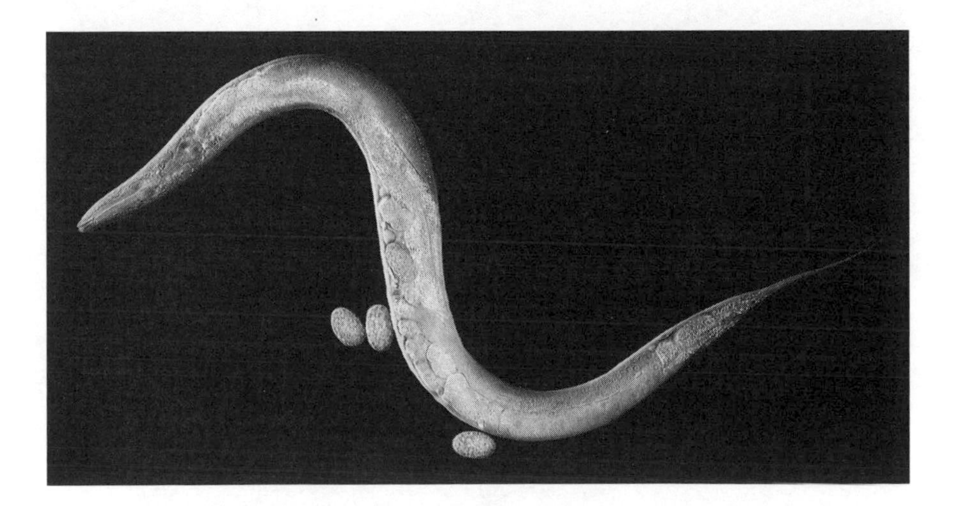

FIGURE 4.5. *C. elegans*

Robert Horvitz shared the 2002 Nobel Prize in physiology/medicine for their work with *C. elegans.*

Another model organism, *Drosophila melanogaster,* is familiar to many of us. Around 1900, Harvard embryology professor William Castle had a graduate student in need of a project. Castle told him to leave out some old grapes for a while and then study whatever showed up. That turned out to be *D. melanogaster*—fruit flies—an organism now studied in labs all around the world. Because of their fulfillment of the model organism's characteristics, fruit flies are used extensively in genetics and developmental biology projects.

Fruit flies have a life cycle of 16 days and produce a new generation every 12 hours. They are fertile, inexpensive, and, in the words of Berkeley geneticist Gerry Rubin, have so much in common with humans that they are "little people with wings." *Drosophila* have 13,600 genes in a DNA of 165 million nucleotide base pairs. They cram all this molecular mechanism into a package about 3 millimeters long, about the size of the V in Venter (more about him later).

Mus musculus (the mouse), long a favorite of medical researchers studying diseases and medications, also fulfills all the requirements of a model organism. In addition, the mouse genome is remarkably similar to the human genome. Genetic comparison studies have already revealed a lot about human structure and function. Further studies should produce additional information.

Other organisms such as the zebrafish, puffer fish, mustard plant (*Arabidopsis thaliana*), and an influenza virus (*Haemophilus influenzae*) have been proposed as model organisms and studied to varying degrees. Model organisms and the machinery needed to carry out studies of them invoke visions of classical descriptive biology, with countless researchers hunched over microscopes or squinting through binoculars on field trips to exotic places where they can view organisms in their natural habitat (think Charles Darwin in the Galapagos).

Physics ↔ Biology ↔ Chemistry

While model organisms are important to biologists, the focus of modern biology has broadened considerably, largely as a result of the influence of several waves of invaders from other disciplines. The activities of these invaders have literally revolutionized the study of biology.

To see how this revolution materialized, let's restate and enlarge the central dogma of molecular biology. Descriptive biology focused on observable traits but found few clues about the molecular machinery associated with these traits. Then chemistry arrived and focused on chemical reactions in living things, which, in turn, began to clarify biological processes. Yet the biggest difficulty remained in the fact that the molecules that drive living systems were too small to show up in any microscope.

The next invasion was that of the physicists, who used X-ray crystallography techniques to map the double helix of DNA. (Think biologist James Watson and physicist Francis Crick, using the data of X-ray crystallographer Rosalind Franklin.) The good news/bad news about DNA was that its overall structure became known, but the details of that structure were too small to see. DNA contains such a huge number of nucleotide base pairs that identifying them and sorting them out presented a difficult problem.

To summarize the position of biology in the 1980s: Molecular biology worked in the region of the very small; classical descriptive biology had staked out that portion of the biosphere large enough to see, including microscopic views. Many of the details in between the microscopic and macroscopic domains of biology were incompletely understood. See Figure 4.6.

Working downward from the large scale to the small proceeded slowly. Studying molecules from the chemical standpoint produced some knowledge, but the pace was that of a snail—which is definitely not a model organism.

In the middle 1980s, a new idea occurred to several biologists. Why not analyze the complete collection of an organism's DNA, referred to as

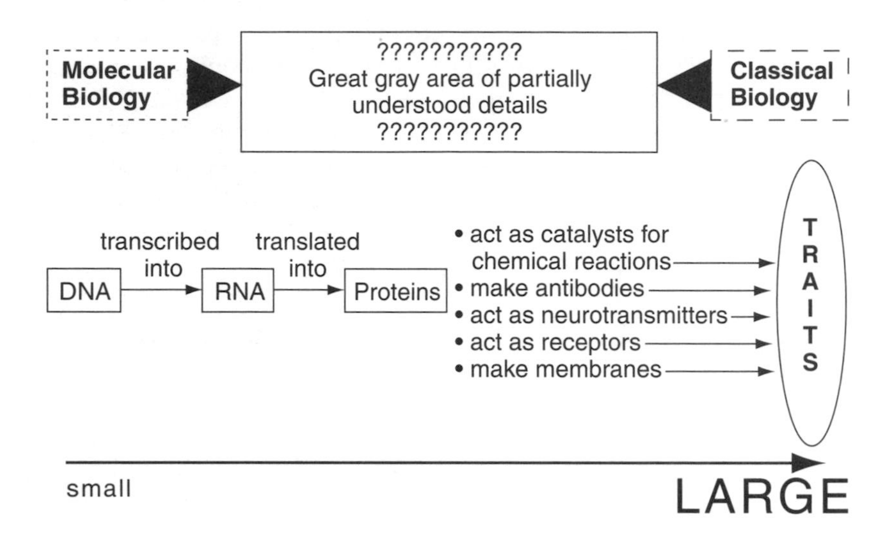

FIGURE 4.6. Biology Summarized

its genome? Better still, make that several model organisms, leading up to the ultimate challenge: the human genome. This bold idea opened the door to the next waves of biology invaders: instrument makers, computer scientists, capitalists, and one especially impatient man, J. Craig Venter.

Mapping the Human Genome: Big Problems Require Big Tools

Before we discuss all the twists and turns that culminated in the successful mapping of model organisms and the human genome, let's explore the details of how the sequence of bases of the jam-packed DNA molecule is ascertained. After all, the human genome consists of 3 *billion* nucleotide base pairs. If you started counting them at the rate of one every second, it would require almost 100 years. Clearly, someone needed to figure out a way to identify them faster. Several techniques had to be perfected to accomplish this feat.

ELECTROPHORESIS In 1937, Swedish biochemist Arne Tiselius developed a method for separating charged particles in suspension on the basis of their relative mass and charge. (See Figure 4.7.) A charged particle in an electric field experiences an attractive electric force that drives it toward the oppositely charged electrode. Because it is immersed in a medium (a gel), the particle also experiences a frictional retarding force, or drag. When the electric force and the retarding force become

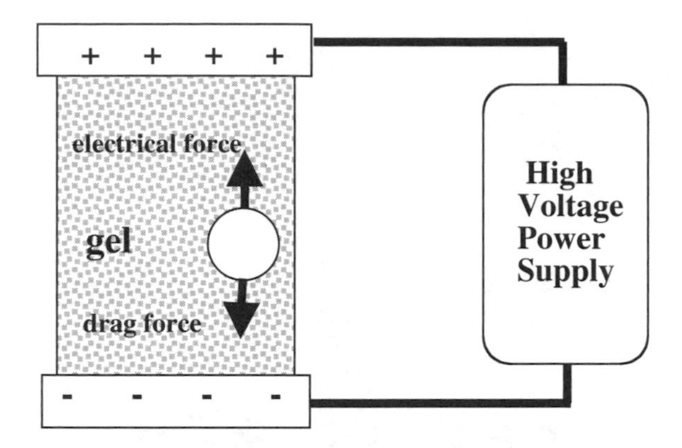

FIGURE 4.7. Electrophoresis Setup

equal, the particle moves at constant speed, referred to as *terminal velocity*. This concept is familiar to sky divers or parachutists, who, because their weight is balanced by the frictional drag force, fall at constant velocity rather than accelerating all the way to the ground.

Tiselius's procedure included a technique for making the resulting pattern visible by using dyes to distinguish the particles from the gel media. He originally applied this process to the separation of proteins in solution, and was awarded the 1948 Nobel Prize in Chemistry for this work. Since then, his procedure has been extended to a variety of particles moving in many different media. Furthermore, several different visualization schemes are now used.

RESTRICTION ENZYMES The development of restriction enzymes began in an unlikely place: research on bacteriophages. Bacteriophages (phages, for short) are viruses that attack bacterial cells by inserting their DNA into the host bacteria, which then reproduces the phage. Phages were originally discovered independently by bacteriologists Frederick Twort (British) in 1915 and Felix D'Herelle (French) in 1917. Research on phages proceeded rapidly because of their potential for killing bacteria that infect humans. Interest eventually waned after the discovery of penicillin and other chemical antibiotics.

Bacteriophages are so prevalent that their number has been estimated to be 10^{30}, which would make their total mass substantially greater than that of humans. They consist almost entirely of proteins and DNA (see Figure 4.8). Since they are viruses, they cannot live independent of a host. Because of their genetic and structural simplicity, they are ideal research subjects for revealing information about their own functioning as well as that of their hosts.

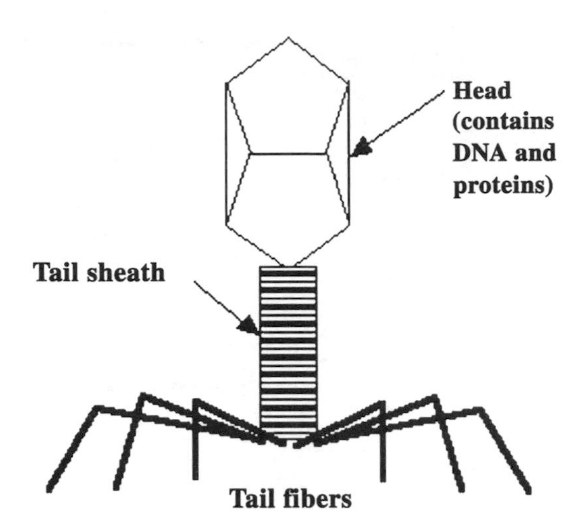

FIGURE 4.8. Bacteriophage

Hamilton O. Smith, a microbiologist at Johns Hopkins, worked with *Haemophilus influenzae Rd* and phage P22 in the late 1960s. By chance, bacteria and phage were incubated together. Smith noticed that the phage DNA's activity diminished over time, which implied that the phage's DNA was being cut by something in the bacteria. Smith and his colleagues isolated and purified the enzyme responsible and identified the way it worked: A protein enzyme within *H. influenzae Rd* cut the phage's DNA by identifying a specific string of six base pairs and cutting the DNA—always at the same location and always in the same way. This enzyme was called a *restriction enzyme.* Besides this enzyme, *H. influenzae* possessed another enzyme, *methylase,* which protected the bacteria's DNA so it didn't get chopped up the same way. The methylase enzyme attached a methyl group to cytosine or adenine nucleotide bases in the bacteria's DNA. Methylation modifies the DNA molecule enough to keep the restriction enzyme from identifying its binding site but doesn't interfere with the bacteria's normal reproduction or metabolism.

Thousands of enzymes have since been discovered that cleave DNA at particular points. Others have been discovered that fasten DNA fragments together. The result of all these discoveries is that molecular biologists now have a toolbox of protein enzymes that enable them to cut or paste DNA at particular nucleotide base pairs.

SANGER DIDEOXY CHAIN TERMINATION METHOD FOR SEQUENCING DNA In 1977, biochemist Fred Sanger of the United Kingdom developed a way to cut a DNA into pieces corresponding to every possi-

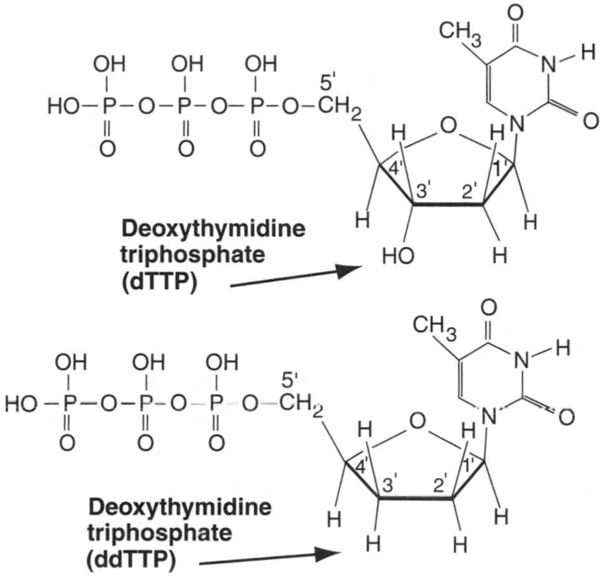

FIGURE 4.9. Deoxythymidine Triphosphate (dTTP) and Dideoxythymidine Triphosphate (ddTTP)

ble length of the original DNA. This technique used a substitute molecule for the normal nucleotide. The substitute didn't form a bond with the next nucleotide in the sequence needed to make the full DNA, so the chain ended there.

Here's an example. In Figure 4.9, the upper molecule has an oxygen attached to the hydrogen at the 3′ position (the carbons are numbered 1′, 2′, 3′, 4′, and 5′), while its hydrogen at the 4′ position lacks an oxygen (thus the prefix deoxy-). The lower molecule lacks an oxygen at the 3′ position as well as the 4′ position, and hence is called dideoxy-. Because of this structural difference, if a DNA molecule is being built and happens to incorporate the dideoxy base, it will not bond with another nucleotide base (at the 5′ location), so the DNA chain terminates at that point. This same process works for the other DNA bases (adenine, guanine, and cytosine), as well. As a result, a variety of lengths of DNA can be generated. (In molecule diagrams, if a ring corner showes no atoms at all, it is understood that this represents a carbon atom.)

The Sanger dideoxy chain termination method for sequencing DNA begins when scientists use restriction enzymes to cut the DNA to be sequenced (the template) into smaller, manageable lengths and heat the DNA until the two strands separate. Then they add dideoxy-base triphosphates to these single-stranded DNA fragments. Next they introduce a DNA polymerase protein enzyme, which begins building copies of the template DNA. Because of the presence of dideoxy bases, the molecules built are not copies of the original DNA but a mixture of every chain length of the DNA. Scientists previously marked the dideoxy-base triphosphates either by a radioactive phosphorus isotope or by an ultraviolet-sensitive dye agent, so the end of each terminated chain is visible.

Then they place this mixture of DNA chains into wells in an electrophoresis gel and turn on the electric field. Shorter DNA lengths encounter less resistance from the medium (usually agarose gel, a substance like Jell-O, except the molecules form additional bonds that crosslink them into a sturdy structure) and therefore travel faster. Often researchers include chains of known length in one of the wells for reference. When the shortest chains reach the end of the medium, the field is turned off. By checking the radioactive or ultraviolet fluorescent markers, scientists can determine the nucleotide base at the end of each molecular chain. Since electrophoresis sorts molecules in order of chain length, simply scanning the electrophoresis results shows the nucleotide base pair order in the template DNA.

Up to the mid-1980s, this technique was used extensively, and many a Ph.D. student completed a dissertation with a multiyear project of sequencing some portion of the DNA of a model organism. Samples had to be extracted from the organism, purified, allowed to react with various chemicals, incubated, set within the gel, and run, and then the data had to be recorded and interpreted. The work was laborious and slow. A typical thesis produced a stretch of 40,000 base pairs of DNA.

Sequencing the Human Genome

Voicing the opinions of many influential biologists, in the March 7, 1986, issue of *Science,* Renato Dulbecco, president of the Salk Institute, called for a massive program to sequence the entire human genome. He argued that just such a major effort was needed to understand the genetic role in cancer. Some biologists, such as Walter Gilbert (of RNA World fame), agreed emphatically. He said, "The total human sequence is the grail of human genetics." (More about that overly strong analogy later.)

Others worried that such a giant project would distort biology beyond all recognition. To sequence 3 billion base pairs using then-current methods would require 10,000 graduate students working nonstop for 15 years at a projected cost of $3 billion. With that level of manpower and funding requirements, there wouldn't be any funding left for other biological projects for 15 years.

One glimmer of hope was the possibility of automated sequencing machines. The Center for Human Genome Research, a unit of the

National Institutes for Health (NIH), officially began operation in October 1990, under the direction of James Watson—yes, *the* James Watson. This project was set up to be a worldwide effort, with most of the work being done by various governmental facilities and universities in the United States and about a third in the United Kingdom, France, Germany, and Japan.

Early efforts focused on developing automated sequencing machinery, which opened the door to biology's next invaders—the instrument makers. Late in 1986, biochemist/M.D. Leroy Hood and biochemist/technologist Michael Hunkapiller formed Applied Biosystems, Inc. (ABI) and developed a machine that could sequence 12,000 nucleotide base pairs a day. By early 1987, a molecular biology lab at NIH headed by physiologist J. Craig Venter tested the ABI 373A Sequencer along with the ABI 800 Catalyst workstation to prepare the samples. Venter's lab sequenced two regions thought likely to contain genes implicated in extremely important hereditary disorders. Although the machines worked extremely well, the specific genes Venter was seeking were not found. In addition, the software identified many false positives that required a lot of manual checking.

Venter was too impatient to wade through lengthy sequences of the genetic alphabet to find the few interesting genes or regions of the genome that code for proteins. He got an idea about how to focus the search. To find the active genes in a particular cell type, he first extracted the RNA from the cell. Since RNA is built from DNA in the first place, it contains the nucleotide base pair sequence from active portions (genes) of the original DNA. Scientists then converted the RNA to more stable DNA (called complementary DNA or cDNA) and attached it to a bacterial chromosome for storage, using the cut-and-paste techniques available through restriction enzymes. Complementary DNA is a standard resource in molecular biology labs all around the world, so its availability is assured. The next step is to sequence the cDNA and compare it to other sequenced genes. This idea, called Expressed Sequence Tags (EST), was not new to Venter. It was first published by chemical biologist Paul Schimmel in 1983 and used extensively by the renowned geneticist Sydney Brenner and others in the late 1980s. But, thanks to the ABI Sequencer and computer workstations, Venter's lab had sequencing capability second to none.

In June 1991, Venter reported in *Science* that, by sequencing ESTs, he had identified about 330 genes active in the human brain. In one stroke, Venter had identified and sequenced more than 10% of the existing world total of known human genes—all in a matter of months. In his direct fashion, Venter pointed out that "improvements in DNA sequencing technologies have now made feasible essentially complete screening of the expressed gene complement of an organism."

Venter's next paper, published in *Nature,* further fueled the immediate negative reaction of some biologists. In that paper, he reported another 2,375 human genes expressed in the brain—double the number of genes sequenced by the rest of the scientific community at the time. Other researchers feared that the cDNAs sequenced by Venter's EST procedure would be funded as a cheaper alternative to sequencing the entire human genome. This approach would miss the subtleties of gene expression such as the *lac* operon, because binding sites for activators and repressors would not be sequenced.

Perils of Patents

A cause of additional trouble was the patenting of ESTs. The NIH Office of Technology Transfer filed a patent application for the first 330+ genes prior to Venter's first publication in *Science* and added 2,421 more genes to the application before the *Nature* article was published. A furor arose quickly and never abated. French research minister Hubert Curien said that "a patent should not be granted for something that is part of our universal heritage." James Watson, director of the Human Genome Project, said that the EST program "could be run by monkeys."

But NIH director Bernadine Healy believed the patent application was appropriate and dismissed the scientists' objections as "a tempest in a teapot." She instructed Watson not to criticize Venter in public and asked Venter to consult with her on human genome research. Watson resigned in April 1992, calling his position "untenable." Meanwhile, Venter had applied for $10 million to expand his sequencing operation, only to have his proposal vigorously rejected by NIH peer review. Venter resigned from NIH in July 1992 and founded The Institute for Genomic Research (TIGR). Starting with 30 ABI 373A Sequencers, 17 ABI Catalyst workstations, and a Sun SPARCenter 2000 computer with relational database software, Venter set out to increase EST sequencing production and sequence genes from model organisms, as well. At a cost of $100,000 per machine, deep pockets were needed to fund the venture.

Funding was accomplished by the next wave of biological invaders: venture capitalists. Wallace Steinberg, head of the HealthCare Investment Corporation and inventor of the Reach toothbrush, put up $70 million to finance the project. Venter could have free rein to pursue his sequencing ideas. Human Genome Sciences (HGS) was set up as a sister company to explore commercial aspects of genomic research. Venter was delighted, proclaiming, "It's every scientist's dream to have a benefactor invest in their ideas, dreams and capabilities." The only catch was that HGS would have 6 to 12 months to review Venter's data before publication. His scientific colleagues were decidedly less enthusiastic. Some even referred to him as "Darth Venter."

Meanwhile, back at NIH, a new director for the National Center for Human Genome Research was announced. The well-respected University of Michigan medical geneticist Francis Collins became the center's second director. As work continued, this publicly funded consortium posted some impressive results. In 1996, the genome of brewer's yeast was completed by over 100 laboratories in Europe, the United States, Canada, and Japan. This eukaryotic single-celled organism contains 6,000 genes out of 12 million nucleotide base pairs in its DNA. Yet, as the halfway point of the Human Genome Project approached, less than 3% of the genome was sequenced, and the public consortium costs were running way over budget. Collins appealed for more speed and novel ideas, but only slight progress resulted.

Shotgun Sequencing

While the public consortium attempted to speed up, Venter's lab, TIGR, tried a totally new tactic: shotgun sequencing. Johns Hopkins researcher and Nobel Prize–winner Hamilton Smith, who had discovered restriction enzymes almost 20 years earlier, had a radical idea: First, shear the DNA into thousands of random-size pieces using sound waves, then sequence the pieces individually using the ABI machines. Store all the sequence data in a computer, and let specially written software find overlaps so the pieces could be stitched together mathematically to form one contiguous DNA. The technique seemed to work in simulations, and Venter didn't shy away from the gamble. TIGR sequenced the entire genome of the *Haemophilus influenzae* bacteria within 13 months, for less than half the sequencing cost of the Human Genome Project. In short order, TIGR completed the sequence of *Mycoplasma genitalium,* the smallest free-living organism known, as well as several archaea genomes. Venter's reputation soared among his fellow researchers as valuable sequencing information was made available for study.

The shotgun sequencing technique worked—for bacteria—but still wasn't fast enough for the Human Genome Project to finish in time. That was soon to change. Late in 1997, the relationship between Venter's TIGR and its sister company, HGS, unraveled completely. Although HGS still owed TIGR $38 million, Venter released the company from the obligation. Venter gained the freedom to release sequencing information faster, without a delay for HGS review.

Venter had even bigger plans, which revolved around Mike Hunkapiller. Since developing the original sequencing machine, the ABI 373A, with Leroy Hood in the late 1980s, Hunkapiller had not only made a number of improvements, he had instituted a substantially changed process. Rather than running DNA fragments down a gel for the electrophoresis portion of the separation, Hunkapiller had developed a technique where the DNA was sent down thin, liquid-filled capillary tubes. With many tubes available in a single run as well as other speed-enhancing improvements, the new machine, the ABI PRISM 3700, was about eight times as fast as existing machines. After showing Venter the prototype, Hunkapiller popped the question: Would Venter team up with him to sequence the entire human genome? After some initial skepticism, Venter agreed. It would take some doing, because the techniques that worked so well on bacteria couldn't be applied directly to the thousand-fold-larger human genome.

Venter relished the challenge. After some initial consultation—more like a warning—with Human Genome Project director Collins, Venter announced the formation of a new company and its main goal: sequence the entire human genome, and accomplish it within three years, substantially before the Human Genome Project schedule. His new company's name: Celera, from the Latin *celeris,* meaning "swift." The company's motto: "Speed matters. Discovery can't wait."

Venter had done it again. The scientific world was sent spinning, but this time Venter's solid record of accomplishment made the critics much more circumspect. Maybe he *could* do it. From Venter's standpoint, it was quite a risk. He had a barely tested prototype sequencing machine and no computer software, because the old methods didn't apply to the new genome. For his next move, Venter opened the door to biology's newest invader: computer programmers, whom he called algorithm scientists. Stitching together overlapping sequences of nucleotide base pairs to create a whole genome was a significant computing problem, but Venter's massive investment in high-end computing equipment and expertise paid off. By 1998, his team had written a program that seemed to work.

As a test, Venter sequenced biology's favorite model organism, *Drosophila melanogaster,* the fruit fly. Both the machines and algorithms functioned well. The 165 million nucleotide base pair, 13,600 gene DNA was sequenced in less than four months, just in time to burn it into CD-ROMs that graced every seat in a scientific meeting the day before the genome paper was released in *Science.*

The Human Genome Project wasn't sitting still through all Venter's maneuvering. With increased funding from many sources, especially the Wellcome Trust in the United Kingdom, the consortium bought new sequencing machines (some from ABI and some from Michael Hunkapiller's competitors) and stepped up its efforts, revising the timetables accordingly. The race was on.

Although the competing parties spoke periodically, tensions were inflamed by the communications media, especially given Venter's direct manner and Collins's personable but firm style. As the end of the race neared, news of the problems in the relationship between the two groups reached the White House. President Bill Clinton told his science adviser, Neal Lane, to "fix it . . . make those guys work together." The job fell to Ari Patrinos, the Department of Energy's genome director. In the spring of 2000, he invited Collins and Venter to his Rockville, Maryland, townhouse for pizza and beer. There they reached a working agreement that allowed the genome sequence announcement of June 26, 2000, to go off without a hitch. In a satellite link-up with Prime Minister Tony Blair of the United Kingdom, President Clinton said, "Modern science has confirmed what we first learned from ancient faiths. The most important fact of life on this earth is our common humanity." (See Figure 4.10.)

The Second-Half Game Plan

Despite the media hype, the race to sequence the human genome was actually a race to a new starting line. To restate the problem: DNA con-

FIGURE 4.10. J. Craig Venter (left), President Bill Clinton, and Francis Collins at the White House on June 26, 2000

tains the plan for an organism's complete function. But before the function can be carried out, the plan must be transcribed into RNA, which in turn is translated into proteins, which then go on to build each individual cell's structure and carry out its functions.

Proteins are the molecules that actually do the work of sustaining life. The genome tells RNA what proteins to build, but variations occur (proteins fold, interact, get sugars or methyls attached, and so on) before they actually carry out their multiple missions, eventually producing recognizable traits. Now you can see why referring to the human genome as similar to the grail may be too simplistic. Knowing the genome sequence doesn't solve everything. Knowledge of the plan is insufficient.

A more appropriate analogy might go like this:

> In the far distant future, an archaeologist discovers a fleet of supersonic transports (SSTs) in a huge cavern, all fueled and ready to go. Test pilots fly the planes, which function beautifully. However, there are no instructions about their use, construction, or history. Teams of engineers put them through their paces, take them apart, and try to reconstruct how they might have been built. There is some progress, but the planes

> are extremely complex and the engineers cannot make sense of many of their functions.
>
> Much later, at an adjacent site, another archaeologist finds a large set of documents written in a long-dead language. There's great excitement that now all the planes' mysteries will be solved. When the language is finally decoded, it is found that the documents are not the plans to build supersonic transports but instead a list of parts needed to make the tools used to build the planes. It's a good start, but a lot more work is needed before the design and all the functioning of the ancient supersonic transport is fully analyzed.

To quote J. Craig Venter, "The sequence is only the beginning." Speaking of his 10 years at NIH researching the protein on the surface of heart cells that senses adrenaline and sets in motion the fight-or-flight reaction, Venter says, "What took me ten years I can now do with a fifteen-second computer search."

Proteomics: The Next Frontier

To solve the problem of the molecular basis of life, first there must be people who know what they are searching for and then are in a position to carry out the search. Here's the essence of the unsolved problem that confronts the people who would solve it: The detailed sequence of nucleotide bases in the human genome specifies the order of amino acids RNA must assemble to complete a particular protein molecule. (An earlier principle said there was a simple relationship between genes and protein molecules: one gene, one protein. It turns out to be more complicated than that. Variations occur between the DNA specification and the eventual proteins that do the cell's work. One gene can ultimately produce many different proteins.) The human genome contains about 30,000 genes—only 2% of the total nucleotide base pairs. The other 98% of the genome is often improperly referred to as "junk" because we don't know exactly what its function might be. The full complement of human proteins encoded by the genome—the *proteome*—numbers well over 100,000, perhaps as many as a million. Proteins are key to the structure and function of a cell. Proteins determine classical biological traits.

The field of *proteomics* focuses on how dynamic networks of proteins control cells and tissues. The next big effort, the next unsolved problem, will be to map the proteome. Here's the game plan:

1. Identify the collective body of proteins built in a particular cell, tissue, or organism.
2. Determine how these proteins interact with other proteins, forming networks.
3. Find the precise three-dimensional structure of each protein that makes it possible for scientists to search for binding sites (for example, binding sites where proteins are most vulnerable to drugs).

While this plan is simple enough to outline, many complicating factors are present. First of all, there is no single human proteome. Brain cells make one set of proteins, fingernails another, blood cells another, and so on. And the proteins built may depend strongly on conditions such as disease, recently consumed substances, and mental or even emotional condition. Each condition or state of an organism generates a different proteome. Besides, proteins are highly complicated structures. They fold themselves in various ways, not necessarily predictably. (You can devote your unused computer power to calculating alternate folding patterns of proteins by accessing the Web site: *folding.stanford.edu/.* (See Idea Folder 8, Protein Folding.)

At an experimental level, large-scale machines automated the capillary electrophoresis process, yielding rapid sequencing of DNA. Several different techniques are currently being used in mapping the proteome, but none has achieved the dominance of the DNA sequencers. Mathematically, the software development also has a ways to go. The difficulties were nicely summarized by a 2001 conference on proteomics entitled "Human Proteome Project: 'Genes Were Easy.' "

The second half of the race has already begun. In January 2002, two groups reported protein interaction maps for *Saccharomyces cerevisae,* brewer's yeast. Additionally, the rough draft of the rice genome was announced on April 5, 2002, by two groups: Beijing Genomics Institute (*indica* subspecies) and the global agribusiness company Syngenta (*japonica* subspecies). The genes of cereal grains have substantial similarities. Several groups that formed part of an international consortium—the International Rice Genome Sequencing Project (IRGSP)—selected rice to study. Rice has 430 million nucleotide base pairs, compared to corn with 3 billion and wheat with 16 billion. All have more genes than humans—possibly leading to gene envy on our part. Sorting out how differences in the genome lead to various organism traits is what will make the second half of this race interesting.

To put the current situation another way, the glass we hope to fill with a complete understanding of how DNA, RNA, and protein molecules work together is both half full and half empty, depending on your perspective.

Implications and Complications

Unlike the other unsolved problems, the proteome hits closer to home. What about us? Human genomes help run each of us, and there are things we'd probably like to identify and adjust. Deciding to apply knowledge goes beyond the considerations of pure science, which is motivated by the curiosity to know. However, it's only human to focus on more practical considerations. The invaders who revolutionized biology weren't driven solely by curiosity. Those who backed the private genome sequencing labs had in mind the good of humankind as well as economic concerns. Once it becomes possible to influence the human condition, another whole set of considerations enters the picture: ethics.

Applying knowledge about the human genome has the potential for great good—and great harm. Perhaps mindful of physics' experience with the Manhattan Project, the first director of the National Center for Human Genome Research, James Watson, devoted 5% of the center's budget to the study of ethical, legal, and social implications of the project. He wrote, "We need no more vivid reminders that science in the wrong hands can do incalculable harm."

While questions about applications of genetic knowledge extend far beyond the scope of this book, let us examine briefly a few of these applications in the hope of clarifying the relevant science so that eventual ethical decisions about their implementation can be better informed.

BIOCHIP Using photolithography techniques similar to those used in making computer chips, hundreds of thousands of biologically active molecules—DNA, RNA, proteins—are deposited in neat rows and columns on a glass or silicon chip. Biological molecules to be tested are treated with fluorescent dye, then washed over the chip. The DNA or protein molecules stuck on the chip could act "as thin strips of molecular Velcro," in the words of inventor Stephen Fodor. The test molecules are complementary to the ones on the chip and would stick and then show up as a fluorescent spot when scanned by a laser. The scanning output would then be displayed and analyzed by a computer and ultimately be used to detect mutations, find information about diseases or treatments, learn which genes interact with which other ones during cell development, and study many other aspects of genetic information.

AGRICULTURAL APPLICATIONS Using restriction enzymes, it is possible to modify the DNA of plants to produce more desirable traits: higher yield, more nutritious food for humans and animals, higher vitamin and mineral content, more disease resistance, increased herbicide resistance so that weeds may be eliminated more easily, more insect resistance, the ability to fix nitrogen so plants become less demanding of soil nutrients, and increased milk production on dairy farms.

HUMAN GENETIC CONTROL Once human traits are fully linked to genes, we may be able to select traits of human offspring and to predict the likelihood of genetic disorders in humans. The ethical consequences of this possibility are far-reaching, to say the least.

STEM CELL RESEARCH In humans, once the egg cell has been fertilized by the sperm, the developing embryo contains all the genetic information to build an entire human being. Cells that have the potential to develop into any cell in the organism are called embryonic stem cells. As the organism develops, cells become specialized and lose the flexibility of stem cells. Stem cells with full capability may be harvested and used for such worthwhile purposes as repairing damaged heart or spine tissue. However, the methods of harvesting such cells have ethical implications that have not yet been worked out fully. An alternative is to wait until a later stage in the development process and harvest adult tissue-specific cells, which may be almost as useful.

CLONING Originally, cloning involved replacement of an egg cell's nucleus with the nucleus from another cell, then implanting the new egg cell into a surrogate mother, who would eventually give birth to an organism whose genetic characteristics are identical to those of the transplanted nucleus. When the techniques became well enough understood, successful cloning experiments were performed on mice, pigs, cattle, and, most famously, the now-deceased sheep Dolly.

So, *could* a human be cloned? Based on current biological practice, it is certainly possible, and some unverified claims have already been made by a group called Raelians. Whether it *should* be done is another question entirely, rooted in ethics and legalities. Of only slightly lesser import is the possibility of using restriction enzymes to cut and paste human DNA into animals, then clone the animals, turning them into factories for the production of medicinal proteins, rare hormones, or even whole organs to be used as replacements for humans with injuries or diseases.

This small sample of applications of genomics (and thus proteomics) gives you an indication of the range of ethical aspects of genomics and proteomics. (For more details, see Idea Folder 9, Genetic Technologies.)

Solving the Puzzle: Why, How, Who and Where, When?

WHY Proteomics holds forth the possibility of developing newer, more effective drugs as well as diagnostic tests. The sheer number of base pairs, genes, and proteins involved, however, poses a huge challenge to those who would analyze, design, and test such applications.

HOW Just look at the numbers: 3 billion nucleotide base pairs, 30,000 genes, hundreds of thousands of proteins in the human organism. They add up to one huge problem, necessitating techniques for dealing with enormous volumes of data. The new subdiscipline of bioinformatics enshrines the latest invaders—algorithm scientists—as legitimate biologists and supplies computing tools to collect, organize, and interpret data of biological significance. Although bioinformatics might be the key to solving the overall problem, the sheer size of the problem may be an indicator of its intractability. At the very least, the complexity of protein interactions may make the overall biological system one in which extremely small variations in inputs, easily supplied by the wide variety of perturbations that normally occur, inexorably lead to radically different outcomes. (Chapter 5 on weather poses a similar problem.)

In some ways, this problem is reminiscent of one in physics in which individual particles are grouped into ensembles whose behavior is predicted on the basis of statistical methods. This approach, called statistical mechanics, works very nicely. In physics, the particles are identical and much more numerous than biological molecules, so the statistics of large numbers lead to convergence. Biological systems deal with nonidentical units, and their number is significantly smaller than, say, the number of atoms in a roomful of air. Thus, the advantages of statistics might not apply to this problem. New types of statistical mathematics may have to be developed. Bioinformatics would be a likely field in which that might occur.

WHO AND WHERE Another possibility is that biology may be assisted by another wave of invaders from some other field or subfield, or perhaps more large project coordinators like Francis Collins or impatient loners like J. Craig Venter. Besides Celera, the list of companies engaged in proteomics research now includes Cellzome in Germany, Hybrigenics in France, and MDS Proteomics in Canada.

WHEN According to professor of pharmaceutical chemistry Alma L. Burlingame of the University of California at San Francisco, "We now have the capacity to identify the protein components of human beings rather rapidly. It's tractable and will occur over the next few years."

Scientists usually make use of two major techniques to find out which proteins are present in selected cells or tissues: two-dimensional gel electrophoresis and mass spectrometry. Several companies are refining and automating these and other techniques as quickly as possible.

With the structure of the human genome established and traits well studied, it might seem that the problem is bounded at both ends, and it's just a matter of applying known principles to work out the details. Wrong. New information about the genome has already yielded many surprises. Don't be surprised if more revelations follow.

CHAPTER FIVE

GEOLOGY

Is Accurate Long-range Weather Forecasting Possible?

Weather forecast for tonight: dark.
—*George Carlin*

Everybody talks about the weather, but nobody does anything about it.
—*Mark Twain*

The study of planet Earth as an entity is the province of geology (literally "Earth science"). The plate tectonics model of Earth describes reasonably well the effects of interactions between the outermost of Earth's four solid and liquid layers. Earth's atmosphere, most notably its weather patterns, however, seem to defy attempts to formulate models that lead to reliable long-range predictions. Since weather is such a prominent feature of our planet, finding a suitable model for weather prediction is the biggest unsolved problem of Earth science.

Earth Weather

Mostly sunny, partly cloudy, occasional showers, possibility of snow, severe weather watch . . . Did you ever wonder if weather forecasters borrowed their equivocal vocabulary from the people who write horoscopes? What about the *Farmers' Almanac,* which forecasts weather years in advance, or Aunt Barbara's trick knee, which always acts up just before a rainstorm?

Weather and its prediction has always been an important practical factor in human survival. The earliest recorded weather references usually were associated with religion or folklore. An Egyptian religion based on sky gods featured rain-making rituals around 3500 B.C.E. Babylonians (3000 B.C.E to 300 B.C.E) connected astronomical bodies with weather events, believing that what looked like a dark halo around the Moon signified the coming of rain. Ancient Greeks produced a series of weather observations and theories, culminating in Aristotle's *Meteorologica* in 340 B.C.E., which collected earlier ideas and fit them into the four-element (earth, water, fire, and air) theory prevalent at the time. After the Scientific Revolution of the 1600s, Aristotle's theories were finally challenged on the basis of experimental evidence, and the global nature of weather and climate was recognized. In modern times, the study of atmospheric conditions has become a legitimate part of the well-respected field of Earth science. Earth's atmosphere is its outermost layer. The four other layers—the inner and outer cores, mantle, and crust—are slowly moving solids and liquids. Since the atmospheric layer is gaseous, it changes most rapidly.

Earth Weather Forecast

Occasionally, Partly, Mostly, or Totally Sunny or Cloudy with varying probabilities of Rain, Snow, Hail, Sleet, Tornado, Hurricane . . .

Temperature: High: 136 °F (58 °C)
Low: −129 °F (−84 °C)
Pressure: 1 atmosphere ± 10%
Humidity: 0 to 100%
Winds: 0 to 231+ mph (possibly higher in tornados)
Visibility: 0 to unlimited
Precipitation: 0 to 523 inches of water per year
Lightning Possibility: Variable

The specific forecast depends on location and time of year.

"NOW HERE'S A SIMPLE FORMULA TO HELP YOU DETERMINE HOW ACCURATE MY PREDICTIONS ARE LIKELY TO BE."

From a scientific method perspective, we've observed the atmosphere for quite a long time and have accumulated quite a large amount of data. Hypotheses to explain the dynamics of the atmosphere utilize well-supported ideas from fluid mechanics, thermodynamics, solar astronomy, chemistry, and other disciplines. So, why do weather predictions, like fish and visits by relatives, become overripe in just a few days? In other words, why is accurate long-term weather forecasting still an unsolved problem?

The answer lies in the size and complexity of the atmosphere. It turns out that after a system reaches a certain level of complexity, mathematical predictions depend so strongly on initial conditions that tiny variations lead to vastly different ultimate results. The short version of this sensitive dependence on initial conditions is called *chaos theory,* which is often mistaken to mean total randomness. Before we explore the details of the problem and the efforts to solve it, we'll survey neighboring planets' weather for comparison, investigate the way Earth's atmosphere developed into its current form, set up the weather forecasters' hypotheses, and trace the origin, development, and relevance of chaos theory. Finally, we'll delve into the question of whether it's theoretically possible to forecast the weather using current mathematical techniques.

Our Neighboring Planets' Weather: The Grass Isn't Always Greener

A planet's atmosphere is the layer of gases that surround the solid (and/or liquid) planetary surface. The atmospheres of the gas giants—Jupiter, Saturn, Uranus, and Neptune—are their most prominent feature. Whatever solid parts these planets have is buried deep under a thick layer of gas. Other bodies in our solar system—Mercury, Pluto, and Earth's Moon—have little or no atmosphere. The remaining three planets—Venus, Earth, and Mars—have situations in between the extremes of too much and too little gas—just right, as Goldilocks would put it.

We'll begin by analyzing weather conditions within the atmospheres of our nearest neighbors, Venus and Mars, and then compare their weather with our experiences on Earth.

VENUS Venus is known as Earth's sister planet because many of its properties are similar to those of Earth. Its diameter is 95% of Earth's, and it has 82% of Earth's mass, 95% of Earth's density, and 91% of Earth's surface gravity. On the other hand, Venus rotates very slowly—243 Earth days for one rotation—and in the opposite direction to Earth's. The axis of Venus's rotation is within 2 degrees of being perpendicular to the plane formed by Venus's orbit, not tilted at a substantial angle (23.5 degrees) like Earth's.

If you found yourself deposited on Venus, it would be familiar in size. The gravitational pull would seem familiar (although the 9% weight loss would be a welcome change for many of us), but there would be no seasons like those on Earth. The planet would be rotating very slowly, but you wouldn't notice it. On Earth, we don't sense rotation either, except when tracking the Sun's path across the sky or observing the periodic darkness (night) that allows us to see the Moon and stars rise and set. Thanks to its thick cloud layer, there is no such possibility on Venus.

Since we've deposited you on Venus, let's temporarily remove its atmosphere and see what would happen. With no atmosphere, the surface temperature would be determined by the interaction between the Sun's rays and Venus's surface materials. According to the principle of conservation of energy, a planet acts like a nonprofit corporation: Income (the Sun's energy arriving at Venus) equals outgo (the energy radiated outward by the planet). Since Venus is closer to the Sun than Earth is, Venus receives more solar radiation than Earth does. The amount of energy radiated away depends on the surface characteristics.

The quantity that measures radiation efficiency is called the albedo. A perfectly reflecting body has an albedo of 1 (think mirror), and a body that reflects no energy at all has an albedo of 0 (think dark asphalt). Without any atmosphere, Venus's albedo would probably be like Mercury's, since their surfaces are similar. Venus's atmosphere-free temperature has been estimated to be approximately 100 °F (38 °C). Although that may seem fairly hot as far as planetary temperatures go, you haven't seen anything yet.

Now let's put Venus's atmosphere back in place. It's approximately 100 times as large as Earth's and consists of 96% carbon dioxide (CO_2), 3% nitrogen (N_2), a little sulfur dioxide (SO_2), and only a trace of oxygen (O_2). At the surface, the pressure is stifling—90 times that of Earth. And that's only part of the story. There are higher-altitude clouds everywhere, but they are pale yellow in color rather than white and gray. That's because they aren't water vapor. These clouds consist of sulfuric acid (H_2SO_4) droplets. Sunlight barely makes it through them. The surface temperature would increase as soon as the atmosphere was put back. It would soar to an incredible high of 870 °F (470 °C).

If you are stuck on Venus and check the weather on VWW (Venus Wide Web) or tune in to Venus TV, you'd get a weather report something like this.

Venus Weather Forecast

Totally Cloudy

Temperature: High: 870 °F (470 °C)
Low: 870 °F (470 °C)
Pressure: 90 times Earth's (90 atmospheres)
Humidity: 0
Winds: less than 3 mph at the surface, 220+ mph at high cloud level
Visibility: Unlimited
Precipitation: Won't reach the ground
Lightning Possibility: Only in clouds

This forecast good for all locations and times because the high winds keep the clouds thoroughly mixed and uniform.

Now let's examine some of the details.

The high and low temperatures on Venus are the same because of the thick atmosphere and its circulation pattern, which distributes energy evenly across the whole planet. The extremely high value of this uniform temperature—hot enough to melt many solids, even metals like lead and zinc—is due to an extremely strong greenhouse effect. (See Idea Folder 10, Greenhouse Gases.) Carbon dioxide in the atmosphere reflects infrared radiation from the planet's surface. This keeps the surface much warmer, a little like the way glass in a greenhouse keeps the plants inside at a higher temperature. Starting in 1961, the Soviet Union explored Venus with 16 spacecraft, called the Venera series, launched over a period of more than 20 years. Spacecraft that landed on Venus functioned for only a short time before being disabled by high temperature and pressure.

The high pressure is a result of the atmosphere's large mass. At Venus's surface, the carbon dioxide is much denser than Earth's air. Trying to walk on Venus would be like wading through a thinned-down version of water, with a pressure corresponding to 3,000 feet (900 meters) underwater on Earth.

There is no humidity because Venus has a minuscule amount of water. One current theory says that Venus started out with plenty of water, but the high temperature resulting from its proximity to the Sun, combined with its developing greenhouse effect, evaporated the water into the atmosphere. There the Sun's photons broke the water molecules down into hydrogen and oxygen. The lighter hydrogen molecules escaped while the chemically active oxygen molecules formed either carbonate rocks (which are still on Venus's surface) or sulfuric acid with sulfur from volcanic emissions. (More about atmospheric dynamic processes in the next section.)

Venus's surface winds move at a very low speed, but in its clouds, 220-mph winds swirl the sulfuric acid droplets all the way around the planet, east to west, within a few Earth days. The mechanism for this circulation is unknown.

Visibility at the surface is unlimited because the atmosphere is clear at the lowest level. Higher atmospheric levels are hazy, and the clouds are almost opaque. The Sun's light filters through the sulfuric acid clouds in a peachy-orange color that is about as bright as a heavily overcast day on Earth.

Sulfuric acid droplets in the clouds collect into large enough drops to produce rain, so there is precipitation. However, it is sulfuric acid rather than water. Since Venus's high temperature evaporates the acid rain, it never hits the ground.

With high winds aloft and clouds rubbing together, there is considerable lightning, but because the clouds are so high—about 50 kilometers or 160,000 feet—the lightning is mostly of the cloud-to-cloud variety rather than the great amount of cloud-to-ground variety we see on Earth.

Venus's weather forecast doesn't include times of sunrise and sunset for several reasons. First, the Sun isn't visible from the surface. Second, Venus rotates on its axis so slowly that it circles the Sun in less time than it takes to make one rotation, so sunrise and sunset wouldn't be as common on Venus as they are on Earth. Further, if the Sun was visible, you'd see it rise in the west and set in the east because Venus rotates in the opposite direction from Earth. If you thought the appearance of the Moon or stars would cheer you up, you can forget that. The clouds wouldn't let any starlight in, and there's no moonlight because Venus has no moon.

To train for a Venus trip, you should set your oven to self-clean to get the temperature to about the right level, then take it down 3,000 feet into the ocean to get the correct pressure. If those conditions aren't disturbing enough, try going without seeing the Sun for long periods of time.

As a tourist attraction, Venus has limited appeal. Nevertheless, the European Space Agency is planning a mission in 2005—the Venus Express—and the Japanese are planning one in 2007.

MARS Our other close neighbor, Mars, has weather that may be more to our liking. Here's what a weather report on MWW (Mars Wide Web) might look like.

Mars Weather Forecast

Mostly sunny

Temperature: High: 81 °F (27 °C)
Low: −207 °F (−133 °C)
Pressure: less than 1% Earth's (0.01 atmosphere)
Humidity: 0
Winds: regularly exceed 100 mph
Visibility: Unlimited, except in dust storms
Precipitation: Carbon dioxide snow near both poles
Dust Storm Possibility: Higher in southern hemisphere summer

This forecast varies with location and time.

The diameter of Mars is 53% of Earth's; the planet has 11% of Earth's mass, 66% of Earth's density, and 38% of Earth's surface gravity. The atmosphere of Mars is 95% carbon dioxide (CO_2), 3% nitrogen (N_2) and almost 2% argon (Ar). The atmospheric pressure at the surface is less than 1% of Earth's, so the total mass of its atmosphere is correspondingly less than 1% of Earth's.

To get a close-up view, let's borrow a trick from Hollywood. Pretend you're suddenly plopped down on Mars and can see for yourself what it's like. The first problem you'd encounter would be the lack of oxygen. Space suits would be required dress on Mars. Even with bulky suits, it would be easier to move around on Mars than on Venus or Earth. The atmosphere is so thin it wouldn't resist any movements. Even better, with the gravitational acceleration only 38% of Earth's, you'd weigh only a little over one-third as much as on Earth. You could jump almost three times as far, and Tiger Woods's golf shots would go out of sight.

The next problem would be the temperature. With all the carbon dioxide, you might suspect there would be a greenhouse effect on Mars similar to that on Venus. There is, but the atmosphere is so much thinner that the temperature isn't affected very much. If we take away Mars's atmosphere as we did to Venus, the average temperature would be −67 °F (−55 °C). Putting the atmosphere back only raises the average temperature to −58 °F (−50 °C). But before you mistakenly leap to the conclusion that Mars is *always* too cold (just as Venus is *always* too hot), be aware that Mars has a few more tricks up its planetary sleeve.

First of all, the length of a Mars day is quite close to that of an Earth day. Mars rotates around its axis in about 24.5 Earth hours. So the pattern of sunrise and sunset would be familiar to an Earth traveler. Also, the rotation axis of Mars is tilted to its orbital plane by about 25 degrees. If you recall, Venus had no such tilt, and that was one of the factors contributing to Venus's bland weather. Similar to Mars, Earth's 23.5-degree tilt causes our seasons because the hemisphere receiving more direct sunlight gets hotter.

During one-half of Mars's orbit around the Sun, sunlight illuminates its southern hemisphere more directly than its northern hemisphere. This is the southern hemisphere's summer and the northern hemisphere's winter. During the other half of its orbit, the situation is reversed, and the northern hemisphere has more direct sunlight, producing summer. Mars adds one more variation to this picture. Its orbit is more elliptical than that of Venus or Earth. In fact, Mars has the third most elliptical orbit among the planets. Because of this, the seasons aren't symmetrical for the northern and southern hemispheres on Mars. During southern summer, Mars is closer to the Sun and receives 40% more sunlight than southern winter. With this in mind, we're now ready to explore Mars's weather.

- *Global atmospheric circulation.* Sunlight heats Martian "air" near the equator, where it rises and moves toward the poles, where it cools and descends. This is similar to the circulation pattern on Earth.
- *Warm and cool "air" masses.* The boundaries of these "air" masses sweep across the Martian landscape just like the weather fronts we know on Earth.
- *More extreme weather in one hemisphere than the other.* Because of Mars's elliptical orbit, the southern hemisphere's temperatures are more extreme than those of the northern hemisphere. Southern summer is hotter, with the temperature rising as high as 81 °F (27 °C); its winter is colder, with the temperature dropping to −207 °F (−133 °C).
- *Dust storms.* When surface winds reach 100 mph, as often happens in southern hemisphere summer, the dust grains (iron oxides or rust) at the surface are picked up and blown around, possibly even expanding to encompass the entire hemisphere or, more rarely, both hemispheres. That's right, dust storms that cover the entire planet have been observed. The details are

incomplete, but computer programs designed to simulate Mars weather indicate that the airborne dust would cool the planetary surface. The concept of nuclear winter was developed by applying similar logic to Earth (more about the Earth versions of these computer programs in the next section). The explosion of large numbers of hydrogen bombs would not only wreak the devastation intended, but the dust thrown into the atmosphere would cool Earth so substantially that it would create extremely cold temperatures, possibly on a global basis. Depending on how fast the dust would settle, a nuclear winter could last much longer than a normal winter.

- *Water ice clouds.* During the northern hemisphere summer, instead of global dust storms, planetwide belts of extremely thin water ice clouds form. These clouds don't extend to nearly as high an altitude as the dust storm particles, and their dynamics are not fully understood.
- *Dry-ice snow.* At both poles, carbon dioxide condenses in the winters to form the solid state directly from the gaseous state. Carbon dioxide in the solid state is called dry ice. Ice-cream vendors and others on Earth use dry ice to keep their products cold but not soggy.
- *Polar ice caps.* The white substance seen in telescopic and satellite photos of the poles of Mars is water ice and dry ice. It is estimated that if all the water ice at the planet's poles melted, it would cover Mars to an estimated depth of 30 feet (9 meters). For unknown reasons, the south polar cap is slightly offset from the geometric pole.
- *Cyclones.* In April 1999, the Hubble Space Telescope found a cyclonic storm system (see Figure 5.1) in the north polar region of Mars. The storm consisted of water ice clouds and was about four times the size of the state of Texas.
- *Sky color.* A clear sky on Mars might be similar in color to Earth's blue, but there's almost always surface winds blowing around some red dust, which makes the sky an orange-brown color that some call butterscotch. Sound delicious?
- *Moons.* Deimos and Phobos are the two moons of Mars. Named for the Greek words for "panic" and "fear," these two moons are small and very rapid in their orbits. Phobos is the closest moon to any planet in our solar system. It zips around Mars almost three times a day. It's tiny and not visible from all spots on the planet, but when it is, viewing it must be a treat because it moves quickly from west to east.

FIGURE 5.1. Cyclone on Mars as Seen by the Hubble Space Telescope

- *Lack of oceans.* Originally scientists thought that Mars's weather was significantly less complex than Earth's, mostly because of Mars lacks liquid-water oceans, which complicate our own weather enormously. Recent missions to Mars have convinced us that the planet's weather is more complex than we thought, with much more variability than expected.

In 2003, the European Space Agency will send a mission to Mars, the Mars Express, which will arrive on December 26, 2003. NASA plans two short-range surface rovers in 2004, a reconnaissance orbiter in 2005, a longer-range rover in 2009, and a soil sample return mission in 2014. We'll learn plenty from them.

If Goldilocks sampled the weather of Venus and Mars before she tried Earth's, she might have said, "The first one was too hot and dense, the next one was too cold and thin, but the third one should be just right." It is.

Home-grown Air

Since the inner planets—Mercury, Venus, Earth, and Mars—are grouped close to the Sun (see Figure 5.2), you might expect them all to be composed of about the same raw material. They are.

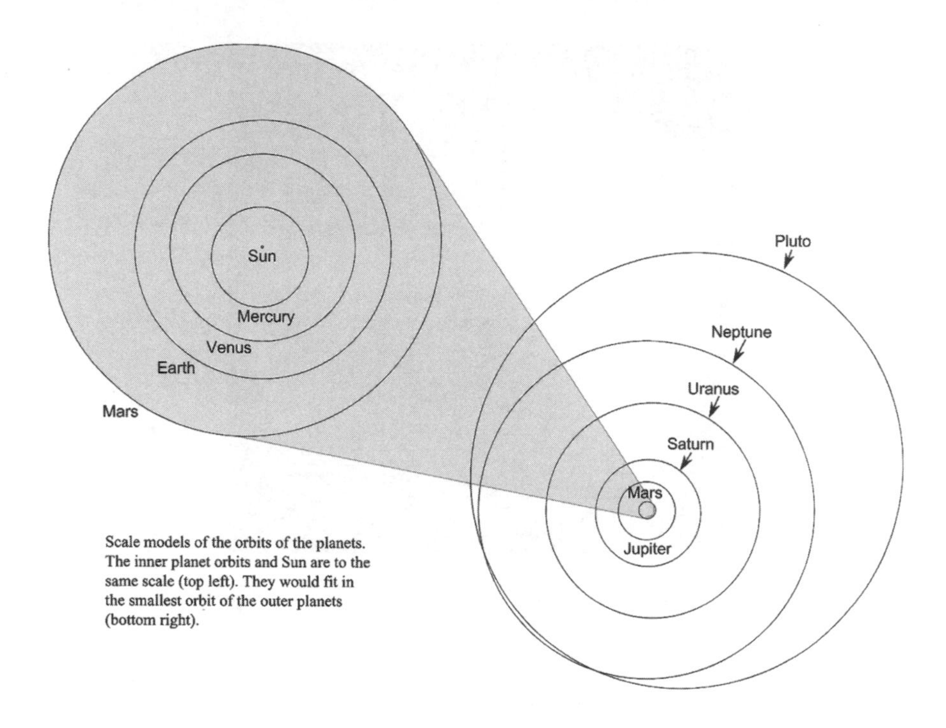

FIGURE 5.2. Solar System Planetary Orbits, Roughly to Scale

As discussed in chapter 3, the planetesimal rain period during the early throes of the solar system's formation deluged all the inner planets with rock and water. Why did Venus and Mars squander their water, while Earth retained its water? To answer this question, we need to look at the processes by which planets acquire gas for their atmospheres and how they can lose that gas.

Gaining Atmospheric Gas

Once the Sun's nuclear furnace fired up, the blast of solar wind (a dilute plasma of mostly protons and electrons, traveling now at about 400 kilometers per second) blew away most of the primordial hydrogen and helium the inner planets had managed to accumulate. Think about poor Mercury. It is so close that every time the Sun sneezes, Mercury says "Gesundheit" as it gets sprayed. About the only gases it has managed to hold on to are ones ejected by the Sun. Over 4 billion years ago, Venus, Earth, and Mars had virtually no atmosphere. Their atmospheres probably formed by three processes: outgassing, evaporation/sublimation, and/or bombardment.

OUTGASSING By the process known as accretion, planetesimals were collected gravitationally, forming the planets. As accretion proceeded, the denser materials sank toward the center of each planet, forming their cores. Less dense rocks didn't sink as far. They formed the mantle of each planet. Chemical reactions in the mantle formed gases that were trapped underground by the pressure of the material on top. Finally, the least dense materials floated to the top and formed the crust. The process of gravitational separation by density is called differentiation. (See Idea Folder 11, Earth: The Inside Story.)

"WOW — WHAT A WASTE OF GEOTHERMAL ENERGY."

As the crust cooled, gas trapped under high pressure in the mantle occasionally burst free, in what is called a volcano. Volcanic eruptions are spectacular events, devastating large areas. For example, Mount Pinatubo in the Philippines erupted in June 1991, spewing 5 billion cubic meters of ash and debris, producing columns 18 kilometers (11 miles) wide at the base and 30 kilometers high. The atmospheric disturbance lowered the average temperature of Earth by 2 °F (1 °C) for about one year.

The surfaces of the three inner planets each show evidence that volcanic activity was far more common in the beginning of their lives. Major products of outgassing are water vapor (H_2O), carbon dioxide (CO_2), nitrogen (N_2), and two sulfur-containing gases: sulfur dioxide (SO_2) and hydrogen sulfide (H_2S)—the familiar rotten-egg smelling gas.

More than 4 billion years ago, volcanic eruptions produced the major atmospheric gases in Earth's first atmosphere. At about the same time, ancient volcanoes on Mars and Venus also produced early atmospheres for those planets.

EVAPORATION/SUBLIMATION Depending on the temperature and pressure at the surface of a planet, liquids may turn into gases (evaporation) or solids may become gases (sublimation). Familiar examples of this process include evaporation of water from puddles and sublimation of solid carbon dioxide (dry ice) into gas, resulting in the smoky special effects often used at theatrical performances. Sublimation is more important on Mars than on Earth because Mars is colder, whereas evaporation is part of a cycle on Earth—a feature that makes our planet unique. On Venus, evaporation keeps the sulfuric acid clouds from raining onto the surface.

BOMBARDMENT Early in the solar system's formation, the solar wind, planetesimals, and comet debris bombarded the inner planets. When they hit, they either carried or generated gas. Although bombardment is a minor contributor to the atmospheres of Venus, Earth, and Mars, it does provide Mercury and the Moon with what little gas they have.

Losing Atmospheric Gas

A planet can lose atmospheric gases in five ways: thermal escape, condensation, bombardment, cratering, and/or chemical reactions.

THERMAL ESCAPE Spacecraft launched from Earth escape in spectacular fashion. Gas molecules also escape, but much less dramatically. First of all, gravitational force exerted by the planet tugs on all the planet's material. Gravitational attraction at a planet's surface is determined by the mass of the planet and its diameter. On each planet, a body must reach a specific speed to escape gravity's clutches.

Planet	Escape Velocity
Mars	5 km/s (11,200 mph)
Venus	10.4 km/s (23,300 mph)
Earth	11.4 km/s (25,500 mph)

Atmospheric gases have a range of speeds, depending on the temperature and the mass of the gas molecules. At higher temperatures, molecules move faster. Lighter molecules move faster than heavy ones. As you can see from the table, Mars would quickly lose light gases such as hydrogen and helium to thermal escape yet could retain more of the heavy gases such as carbon dioxide. Venus and Earth, with their higher escape speeds, retain gases more easily.

CONDENSATION Just as high temperatures cause liquids to evaporate and some solids to sublime, the reverse process also occurs: at cool temperatures, atmospheric gases can condense to form liquids or even solids. Probably the best example of this occurs on Mars, where carbon dioxide near the poles condenses during Martian winter to form solid carbon dioxide, or dry ice. Condensation even occurs on Earth's Moon. In 1998, the *Lunar Prospector* orbiter discovered water ice in deep craters near both lunar poles. The ice probably came from comet tails impacting the Moon, which may have condensed onto the lunar surface in areas not in direct sunlight. The ice may have arrived billions of years ago and remains there still.

BOMBARDMENT Bombardment may generate an atmosphere for a planet that starts out with almost none. It also can cause a planet that does have an atmosphere to lose gases. The solar wind can give upper-atmosphere gases energy to escape. Solar photons can break molecules apart into smaller ones (by a process called dissociation), which then escape because of their lesser mass.

CRATERING Larger bodies impacting on planetary surfaces also can transfer enough energy to gas molecules to allow them to escape. Smaller planets with a lower escape velocity are particularly vulnerable to this process.

CHEMICAL REACTIONS Depending on the chemical activity of the molecules involved, reactions between gases and surface rocks or liquids may occur, taking atmospheric gases out of circulation. Chemical reactions early in our planetary history tied up much of Earth's carbon dioxide into the form of limestone rocks, thereby removing significant amounts of carbon dioxide from the air.

Gaining or Losing Atmospheric Gas

Let's apply these principles to the inner planets to see how their initial atmospheres evolved into today's versions. We'll start with Venus and Mars and save Earth for last.

Venus

Process	Action	Comment
GAINING GAS		
Outgassing	Produced plenty of carbon dioxide, water, nitrogen, and sulfur compounds	Larger planet produced more gas
Evaporation	The high temperature evaporated most of the planet's water into vapor early in the atmosphere's development	*Positive feedback:* Water vapor and carbon dioxide produced the green house effect, heating the atmosphere, evaporating more water . . .
Bombardment	Minor effect	
LOSING GAS		
Thermal escape	Hydrogen atoms released from water molecules by high-energy photons; other light gases escape because of high temperatures	
Condensation	Small effect	Surface too warm for gases to condense
Bombardment	Solar wind knocked off oxygen in upper atmosphere	
Atmospheric cratering	Minor effect	
Chemical reactions	Oxygen atoms released from water molecules by high-energy photons were tied up in reactions with surface rocks	

Mars

Process	Action	Comment
GAINING GAS		
Outgassing	Produced plenty of carbon dioxide, water, nitrogen, and sulfur compounds; originally thick atmosphere	Mars's smaller size produced less gas, and the planet cooled faster
Evaporation	Some water evaporated, but not as much as on Venus or Earth	Mars had substantial liquid water at first
Bombardment	Minor effect	
LOSING GAS		
Thermal escape	Hydrogen atoms released from water molecules by high-energy photons and other light gases escape because of high temperatures	
Condensation	Water condensed, forming liquid; as the temperature fell, the liquid water turned to ice, and the carbon dioxide condensed, forming dry ice	*Negative feedback:* As gas condensed, the greenhouse effect lessened, cooling the atmosphere, allowing more gas to condense . . .
Bombardment	Solar wind knocked off oxygen in in upper atmosphere	
Atmospheric cratering	Minor effect	
Chemical reactions	Oxygen atoms released from water molecules by high-energy photons were tied up in reactions with surface rocks: reddish-brown iron oxide (rust) makes Mars the red planet; carbon dioxide was locked up in carbonate rocks	

The major difference between our neighboring planets and us is water. Venus's water evaporated because of the planet's high temperature. The evaporation helped produce the runaway greenhouse effect, and then was lost because solar photons broke up the water molecules into hydrogen and oxygen. Mars's water sloshed around the surface in liquid form for a while. The greenhouse effect, however, wasn't strong enough on Mars to keep the water in the gaseous state, so it condensed. The temperature dropped further. Water turned into ice, which still sits, mostly below ground, at the poles.

Now, to Earth. We know what the final result for our water is: We've still got it, in all three forms: vapor, liquid, and solid. Not only does water distinguish Earth from our neighboring planets, it helps give us the variable weather that we cannot quite predict.

Earth versus Mars and Venus

Let's compare Earth with Venus and Mars to see how Earth's weather conditions originated.

Property	Earth	Venus	Mars
Diameter	100%	95%	53%
Mass	100%	82%	11%
Density	100%	95%	66%
Surface gravity	100%	91%	38%
Time for one rotation	24 hours	243 days	24.5 hours
Axis tilt	23.5 degrees	2 degrees	25 degrees
Albedo	0.36	0.72	0.25
Temperature *without* atmosphere	−9 °F (−23 °C)	100 °F (38 °C)	−67 °F (−55 °C)
Temperature *with* atmosphere	59 °F (15 °C)	870 °F (470 °C)	−58 °F (−50 °C)
Atmospheric contents	77% nitrogen, 21% oxygen	96% carbon dioxide, 3% nitrogen	95% carbon dioxide, 3% nitrogen

Since all three planets started out with similar atmospheres, generated by volcanic outgassing of mostly carbon dioxide and water, we need to answer several questions.

WHY DID EARTH MANAGE TO RETAIN ITS WATER, WHILE VENUS AND MARS LOST THEIRS? We've already seen how Venus and Mars lost their water: Venus was too hot; Mars was too cold. On Earth, water participates in several cycles, the most notable of which is the familiar hydrologic cycle, in which water evaporates from oceans, is carried by winds to land, precipitates in the form of rain or snow (some falls on the ocean), drains back into the ocean, and starts evaporating all over again. The cycle is driven not only by moderate temperatures but also by atmospheric circulation patterns, in turn assisted by the planet's axis tilt and rotation rate.

WHAT HAPPENED TO EARTH'S CARBON DIOXIDE? Earth's carbon dioxide wasn't lost; it's just hidden, thanks to the action of liquid water. Carbon dioxide in the air dissolves in the ocean. There it reacts with silicate minerals to form carbonate rocks, which fall to the ocean floor. That's where the carbon dioxide goes. It doesn't remain there, however, because this step is part of a cycle. The plates of Earth's crust move because of currents in the underlying mantle. Carbonate rock gets conveyed into the mantle, where, eventually, the rock is heated. The carbon dioxide escapes into the atmosphere as a result of volcanic eruptions. Once in the atmosphere, it can again dissolve in the oceans and . . . By the way, if you're wondering how the silicate minerals got into the ocean, they were eroded from the land by rainwater. This entire process is called the *carbonate-silicate cycle*. Because this cycle requires liquid water, it can function only on Earth.

WHERE DID EARTH GET ITS OXYGEN? The abundance of oxygen in Earth's atmosphere has a simple source: life. The complete story, however, is a bit more complex. Once living organisms developed, it wasn't long before one life-form used energy from the Sun to build complex hydrocarbon molecules from readily available molecules of water and carbon dioxide. This process of photosynthesis appears to have started early in life's history and pumped out oxygen as a by-product.

Oxygen is highly reactive chemically, so for approximately 2 billion years after photosynthesis started, the oxygen by-product simply reacted chemically with surface rocks. Only when surface rocks were fully oxidized did oxygen start to accumulate in the atmosphere. Once it started piling up, two things began to happen. First, the oxygen that rose to higher levels of the atmosphere was torn apart by solar photons. This tearing apart of oxygen molecules resulted in the formation of a new and unstable molecule called ozone (O_3). Ozone would be little more than a chemical oddity except for its ability to absorb ultraviolet radiation. Once a sufficient amount of ozone accumulated in the upper atmosphere, it served to shield Earth's surface from life-threatening ultraviolet radiation. It then became possible for life to move onto land and breathe oxygen, a combination that made possible new life-forms. For example, us.

The impact of life on the atmosphere began early, and continues. The most recent controversy has to do with civilization's addition of carbon dioxide to the atmosphere and the possible effect on greenhouse gases. (See Idea Folder 10, Greenhouse Gases.)

Earth's atmosphere is the raw material from which weather forecasters make their fearless predictions. It is quite different from that of our neighboring planets and has built-in complexities that make forecasting difficult and challenging. Predicting the weather turns out to be much trickier than it would appear.

Weather and Climate: Hypotheses (Pretty Good); Predictions (Not So Good)

With this unique collection of atmospheric gases on Earth serving as observations, the next step in the scientific method is to generate a hypothetical model of how the atmosphere functions in both the long term (climate) and the short term (weather). Thanks to the efforts of Isaac Newton in the 1600s, the motion of bodies is well described by a set of equations that are general and powerful. In fact, much of the scientific activity of the 1700s and 1800s applied Newton's ideas to different cases, such as large bodies, small bodies, liquids, and gases.

One of the beauties of Newton's laws was that once conditions were specified at one particular time, all future states of motion could be calculated. Philosophically, this was referred to as determinism. The power of this method was enormous. Accurate calculations could be made for the position of planets, tides could be forecast years in advance, and the paths of cannonballs could be plotted. In addition, such forecasts could be run backward as well as forward, enabling analysis of the past as well as the future.

One of the implications of determinism is that future behavior of a system can be predicted simply by specifying the conditions of the system at some earlier time. These earlier conditions are referred to as initial conditions. Figure 5.3 shows a simplified version of this process; the graph could represent the distance a cannonball would travel as a function of the angle at which it is fired. If the angle is set within a range of several degrees, the range could vary substantially. If more accuracy is required, the angle must be set within a smaller range.

In principle, results as accurate as desired could be obtained simply by making the initial conditions correspondingly accurate. Implicit in the measurement that tests a prediction is the idea that increasing the accuracy of measurements would improve the accuracy of predicted results. Although this assumption went unchallenged for a long time, in the late nineteenth century it was questioned—under odd circumstances.

In 1887, King Oscar II of Sweden offered a cash prize to anyone who could show mathematically that the orbits of the planets in the solar

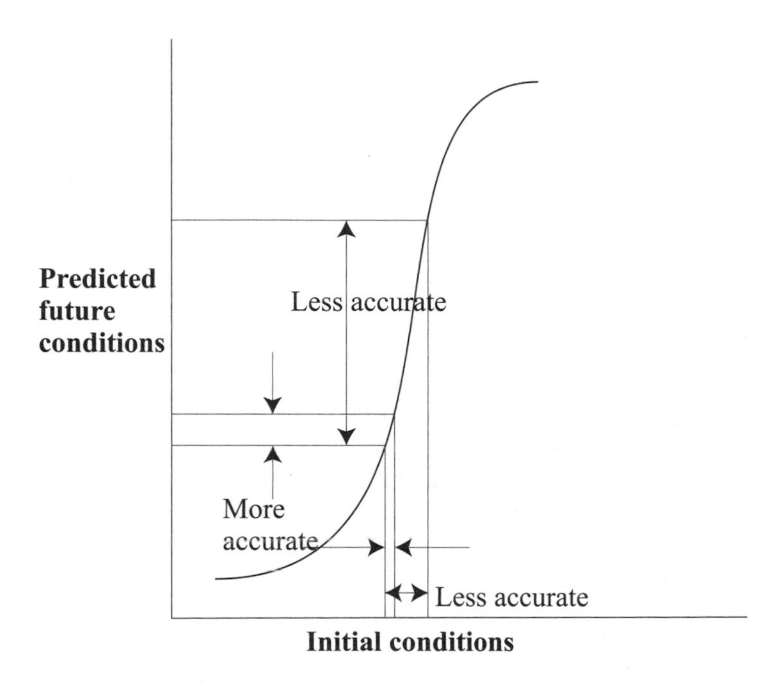

FIGURE 5.3. Accuracy of Predicted Future Conditions Depends on Accuracy of Initial Conditions

system were stable. The winner, Jules Henri Poincaré, didn't solve the problem fully, but did well enough to win the prize. He published a paper entitled "On the Problem of Three Bodies and the Equations of Equilibrium." Poincaré came across the unusual situation that "small differences in the initial conditions produce very great ones in the final phenomena." Since he was a first-rate mathematician, Poincaré was able to show that when systems reach a certain level of complexity, achieving precise results would require infinitely accurate initial conditions. For some time, Poincaré's ideas were regarded as little more than a mathematical oddity. As we'll see shortly, 70 years later, they returned with a vengeance.

Meanwhile, back to weather forecasts. An interesting development occurred during World War I. Lewis Fry Richardson had worked at a variety of scientific endeavors, including the British Meterological Office. When war broke out, he found a way to assist the effort without violating his pacifistic beliefs: He drove an ambulance in France. In his spare time, he devised a mathematical model for weather forecasting based on dividing Earth's surface into cells, obtaining weather data for each cell, and using a mathematical technique called finite difference analysis to predict subsequent weather. His model never worked successfully, but he did publish the technique in 1922 in a famous paper titled "Weather Prediction by Numerical Process." Richardson attributed the model's failure to insufficient data and the difficulty of having to carry out laborious computations manually.

Before long, hand calculations took a backseat to computers. By 1953, Princeton-based Hungarian mathematician John von Neumann had used ENIAC (Electronic Numerical Integrator And Computer) at Princeton University on many interesting problems, including Richardson's equations. Although this early computer work allowed moderately successful weather predictions to be made, there was plenty of room for improvement.

The computer represented a very useful new tool. In 1960, Edward Lorenz had a new computer delivered to his office. Lorenz had studied mathematics at Harvard and now taught meteorology at the Massachusetts Institute of Technology. As a test for the computer, Lorenz programmed the 12 nonlinear equations that governed fluid flow and applied them to weather. These equations included the effects of pressure, wind velocities, air temperatures, and humidity. By modern standards, Lorenz's computer was quite primitive. It did, however, produce reasonable-looking results.

One particular run seemed so interesting that Lorenz decided to extend it. Because computers were slow in those days, he started the run in the middle and typed in a number, which happened to be 0.506, from the printout. Then he went for coffee while the computer churned away. Upon his return, he was shocked to find that the part of the new run that overlapped the old one didn't match the previous results. It wasn't even close. After a good bit of checking, Lorenz found that the machine used six decimal places internally but only reported three. The printout's 0.506 was really 0.506127 inside the machine.

But how could that small difference in input produce such a huge divergence in output? Edward Lorenz had rediscovered Poincaré's ideas. In his 1963 paper, "Deterministic Non-periodic Flow," Lorenz pointed out how sensitive later results are to initial conditions.

Figure 5.4 is a plot of a three-dimensional function generated by nonlinear equations of this type. Note that although the values never converge to a single point, they oscillate around two points as if they were attracting the function, hence the name "strange attractor."

To dramatize the point about the small differences leading to large consequences, and perhaps guided by the visual image of the strange attractor, one of Lorenz's subsequent papers was titled "Does the Flap of a Butterfly's Wings in Brazil Set Off a Tornado in Texas?" This "butterfly

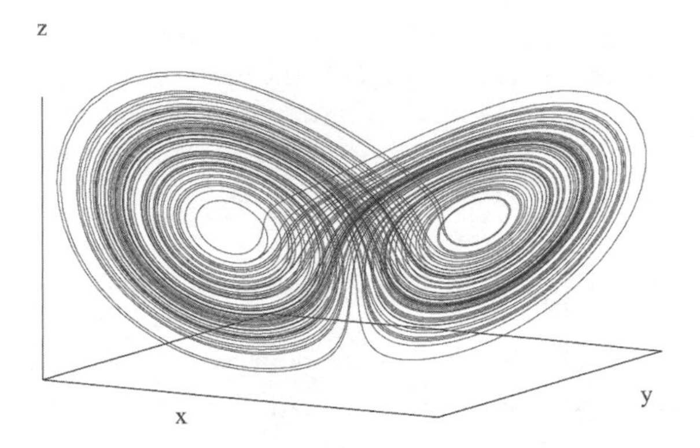

FIGURE 5.4. Strange Attractor

effect" was soon generalized. Systems of equations that exhibited this behavior were generated and studied without regard to their applicability to physical systems.

Eventually, a whole new field of mathematics sprang up, with the possibly misleading title of chaos theory, as named by mathematician/scientist James Yorke from the University of Maryland. (See Idea Folder 12, Chaos Theory.) Unfortunately, the word *chaos* implies complete ran-

domness, which is not truly the case. Weather isn't totally random. The overall weather picture is well known: Summer will be warm and winter will be cool. What escapes us are the details of just how warm or just how cool the weather will be, and whether extreme weather conditions will occur weeks or even several hours hence.

Solving the Puzzle: How and Where?

There are several schools of thought about ways to achieve more accurate long-range weather forecasts.

HOW AND WHERE

Improve Current Techniques

- *Upgrade the observations of weather information.* More and better data are needed. There are areas on Earth where extremely little information is gathered, especially mountainous regions and over the oceans. Two large-scale ocean-current flows, EI Niño and La Niña, produce large-scale weather patterns that have significant impact on world weather, especially affecting agriculture. Accurate long-range forecasts can save farmers hundreds of millions of dollars. Projects such as ARGO, part of the Global Climate Observation System, are placing 3,000 free-floating observatories in the world's oceans to monitor weather and water conditions.
- *Improve the quality of the modeling.* While mathematical modeling is considerably more sophisticated than that used by Edward Lorenz in the 1960s, there is room for improvement. Some of the physical processes that govern weather are quite complicated. Topography and soil characteristics must be accounted for, and ocean and cloud dynamic movements must be taken into account. Current models merely approximate highly complex processes in order to speed the calculation and keep computer memory demands within present capabilities. Further, different agencies maintain several different models, each using a variety of approximations.
- *Reduce the model grid size.* Early global weather forecasting models used a grid with individual elements hundreds of kilometers in size. Today's grid size has been reduced to tens of kilometers, with less than five kilometers a near-term goal. Over a smaller area, more accurate modeling can occur, but providing such accuracy requires supercomputers. (Recall biology's need for massive computer power, referred to as bioinformatics.) Two fundamentally different philosophies of how to build supercomputers exist: massively parallel processing and vector

calculations. Massively parallel processors connect large numbers of general-purpose processors, each of which performs some portion of a complex calculation, and the individual results are integrated. Vector processing uses specialized microprocessors, designed specifically to solve particular problems most efficiently. In the past, U.S. computer designer Seymour Cray built extremely fast supercomputers based on vector processing. While his technique has fallen out of fashion in this country, it has been adopted by the Japanese company NEC. Rather than work with a reduced grid for the whole Earth, it has been suggested that the forecasting quality of global models would be improved by focusing on variable grid sizes in particularly critical areas.

- *Ensemble forecasting.* Ensemble forecasting is a technique that recognizes the sensitivity of models to small changes in initial conditions. The idea is to run the model several times, using different initial conditions, and see how the results vary. If rain results in 4 runs out of 10, for example, a forecaster can say there is a 40% chance of rain. Typically, models are run more than 10 times—often 17 runs, and sometimes as many as 46. A variation on this technique is to compare the results of several different models and make the forecast based on a weighted average. The computer is used as an aid by experienced meteorologists, who examine the results and sometimes override a computer-generated forecast based on their experience.

Acknowledge the Impossibility of Detailed Long-range Forecasts and Analyze Only General Trends

As science writer James Gleick put it in his book *Chaos: Making a New Science,*

> suppose the Earth could be covered with sensors spaced one foot apart, rising at one-foot intervals all the way to the top of the atmosphere. Suppose every sensor gives perfectly accurate readings of temperature, pressure, humidity, and any other quantity a meteorologist would want. Precisely at noon an infinitely powerful computer takes all the data and calculates what will happen at each point at 12:01, then 12:02, then 12:03 . . . The computer will still be unable to predict whether Princeton, New Jersey, will have sun or rain on a day one month away.

The well-established weather forecasting infrastructure regards forecasting impossibility as unacceptable. Until more accurate forecasts of longer than two weeks' duration are demonstrated, the fundamental unpredictability of weather must be taken as a real possibility. In some respects, this is similar to another Earth science prediction problem: earthquakes. (See Idea Folder 13, Earthquake Prediction.)

Develop a Totally New Approach

While chaos theory and *catastrophe theory,* which deals with sudden shifts from one pattern to another, have many interesting characteristics as purely mathematical entities, they need to be better integrated with physical reality to be useful in a scientific sense. A fresh approach based on simple rules developed in computer programming has been published by Stephen Wolfram in a book titled *A New Kind of Science.* His ideas might have an impact on weather forecasting and other areas of science, but a great deal of work will be required to integrate his abstract mathematical modeling techniques with the reality of the universe.

Currently a project called *climateprediction.com* runs atmospheric models on home computers in a background mode as screen savers. This massively parallel computing effort is similar to the SETIathome and Folding@Home projects discussed in Idea Folders 4 and 8, respectively. Sophisticated atmospheric models are run for a wide variety of initial conditions to predict the weather and climate in 2050. The predictions will be compared to the actual 2050 conditions and perhaps shed light on the modeling. Tens of thousands of people have already volunteered their computers for a reward that is purely altruistic.

This project's aim is expressed in a quote that captures the spirit of weather forecasting:

> Helps one to explain the Past, which in turn,
> Helps one to understand the Present, and thus,
> To predict the Future, which leads to
> More influence over future events, and
> Less disturbance from the Unexpected.
>
> —*Charles Handy*

CHAPTER SIX

ASTRONOMY

Why Is the Universe Expanding Faster and Faster?

> This is the exploration that awaits you! Not mapping stars and studying nebulae, but charting the unknown possibilities of existence.
>
> —*Q to Captain Picard,* Star Trek

Astronomy or, more precisely, cosmology studies the origin, evolution, and large-scale structure and dynamics of the universe. Until recently, astronomy/cosmology's biggest unsolved problem was a dynamic one, determining whether the universe will expand forever or eventually contract. The discovery of the universe's increasing rate of expansion, which indicates that the universe will expand forever, may have solved that problem but created another one. The cause of this increased expansion, something called dark energy, seems to run counter to the current understanding of the forces that determine the dynamics of the universe. Finding an explanation for dark energy has become astronomy's biggest unsolved problem.

Contents of the Universe

"What's out there?" is the classic question people ask when they look into the sky.

Astronomy's attempts to answer this question for the whole universe have tantalized us with fascinating answers and frustrated us with fascinating questions.

The overall contents of the universe can be summarized in terms of its mass/energy. (Mass and energy were recognized as interconvertible through Albert Einstein's famous equation relating them: Energy = (mass)(speed of light)^2, or $E = mc^2$). The following table presents the latest estimates for the mass/energy contents of the entire universe, along with brief comments regarding what we know about each.

Universe Constituent	% of the Universe's Mass/Energy	Comments
Dark energy	73	Causes faster and faster expansion of the universe. Although unseen and of an unknown nature, the powerful effects of dark energy have been noted.
Dark matter	23	Also not yet observed, dark matter accounts for rapid rotation of galaxies and clusters of galaxies.
Ordinary matter	4	The observed bright stars, galaxies, and clusters of galaxies.
Neutrinos	<1	An upper limit for their total mass has been set, but the actual value is not yet determined.

The implications of this inventory are startling: *Although totally undetected, dark energy and dark matter are claimed to constitute 96% of the universe and dominate its motion.*

It's legitimate to wonder how astronomy managed to arrive at this current understanding. Like a good detective story, our understanding was painstakingly constructed, one scientific method step at a time. Nowadays, the sequence typically works like this: An improved or new piece of experimental apparatus enables something new to be observed. Theorists then attempt to explain the new information with existing theories or they develop alternate hypotheses. Predictions are then made, and new experiments are performed to see if reality matches the forecast. (Think observers gleefully handing hot potatoes to theorists.)

In this chapter, we'll describe how astronomy's understanding has been assembled. We'll focus particularly on the groupings of stars called galaxies and the techniques used to measure distances to stars and galaxies as well as their speeds. Finally, we'll explore the road to possible solutions to the problems associated with dark energy and dark matter dominating the universe.

Measuring Stellar Distances

The universe is filled with an extraordinarily large collection of objects (think billions and billions, as some of you may remember hearing astronomer Carl Sagan say). Let's begin with a seemingly simple question about one of these objects, a star. How far away is a particular star?

As we look at stars overhead, our usual sense of depth perception fails us. They all look the same distance away—far. Planets and stars are so far away that they all seem to be located at the same distance. That's why the sky looks like a dome over our heads.

Because our two eyes look at an object from slightly different positions, each eye looks along its own sight line. This phenomenon is called parallax, and surveyors use it to make accurate distance determinations. Because of the small separation between our eyes, they cannot be used to judge long distances very accurately.

It may seem surprising, then, that astronomy's simplest technique for determining distances to celestial objects is based on parallax. Here's how it works. If the same star is observed at the beginning and end of a six-month interval, it is seen along two different sight lines (just as our eyes see a distant object from two different perspectives). (See Figure 6.1.) Measuring the angle between these sight lines (the parallax angle) and knowing that the baseline of the triangle is the diameter of Earth's orbit around the Sun enables us to calculate the distance to the star using trigonometry. This feat was first accomplished by German astronomer Friedrich Bessel in 1838, when he measured the distance to the star named 61 Cygni.

This distance-measuring technique forms the basis for defining the unit of distance most often used in astronomy, the *parsec,* abbreviated pc. A star whose parallax angle is 1 arc second (60 seconds in 1 minute,

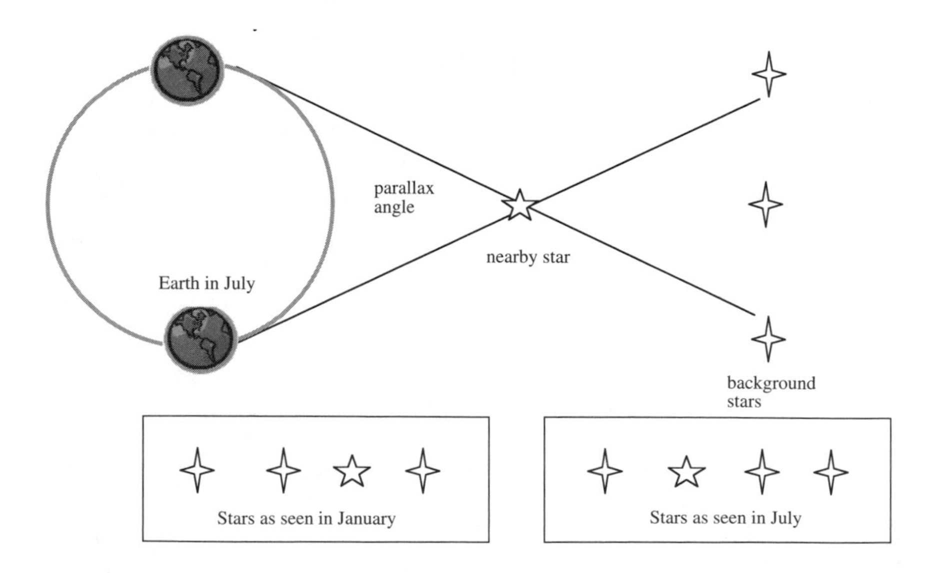

FIGURE 6.1. Distance Measurements Based on Parallax

60 minutes in 1 degree, 360 degrees in a whole circle) when sighted six months apart is defined to be 1 parsec away. Our nearest star, Alpha Centauri (actually a system of three stars), is slightly more than 1 parsec distant. If you flew to Alpha Centauri at supersonic speed, the trip would require more than 1 million years. Even light, with its dazzling speed, requires more than four years to make the trip.

There are more than 300 stars within 10 parsecs of Earth, so we can obtain the distances to these closest neighbors by parallax. Because stars located farther away have ever-smaller parallax angles, a limit is reached at about 100 parsecs, beyond which the angle is too small to measure accurately. Thus, stars and galaxies at thousands of parsecs (called kiloparsecs, kpc) or millions of parsecs (called Megaparsecs, Mpc) are too far away to measure their distances using parallax. To deal with this problem, other techniques, which we will examine later, were developed.

Galaxies—Early Theories and Observations

Next, let's track the development of astronomy's understanding of galaxies. The term *galaxy* is Greek for "Milky Way." Swedish philosopher Emanuel Swedenborg theorized that all the stars formed one large group, with the solar system just one part. In his book, *Principia* (1734), he proposed that our solar system of Sun and planets was formed from a rapidly rotating nebula. The source of Swedenborg's information wasn't any scientific observation, although he did study science. He got his information from a séance that allegedly included visitors from heaven. Later visions encouraged Swedenborg to reveal theological information, and a religion (the "New Church") eventually sprang up from his teachings.

The galaxy story continues with an Englishman, Thomas Wright of Durham, who built scientific instruments and model solar systems that he sold to the nobility. In his 1750 book, *An Original Theory of the Universe,* Wright proposed that stars in the Milky Way are distributed in a kind of shell or disk. He declared, "I can never look upon the stars without wondering why the whole world does not become astronomers." As a scientific instrument maker, he undoubtedly had access to telescopes. However, he published no astronomical observations. Wright's book also dealt with religious matters, such as the physical location of God's throne.

A review of Wright's book in a Hamburg journal caught the eye of the brilliant philosopher Immanuel Kant. Although Kant misread the account of Wright's work, he proceeded to extend it in a constructive direction. In 1755, Kant proposed that the Milky Way was a lens-shaped disk of stars, rotating about its center. Further, he suggested that the fuzzy patches of light called nebulae were actually systems of stars similar to the Milky Way, but very far away. Kant referred to them as "island universes." At the time, there was no way to estimate the distance to these nebulae. Even Bessel's parallax technique, developed almost 100 years later, couldn't be extended that far.

So, the beginnings of astronomy's analysis of galaxies came from a theologically oriented instrument maker and a philosopher. A scientific observer was the next major contributor to the understanding of galaxies. Interestingly, he wasn't looking for galaxies; he was compiling a list of objects to avoid when searching for comets. Charles Messier (1730–1817) was such an avid comet hunter that he was called a "comet ferret" by King Louis XV of France. In his viewing lifetime, Messier discovered or independently codiscovered 20 comets and observed 24 others. In his searches, he often found objects that did not move and were therefore not comets. With the small telescopes Messier used—they were less than 3.5 inches in diameter—individual stars within nebulae could not be resolved. The "nebulae" he observed were little more than blobs of light of unknown origin. He compiled a list of the locations of over 100 nebulae and assigned them numbers. For example, M31 is now known as the Andromeda galaxy, and M100, shown in Figure 6.2, is the Whirlpool galaxy.

Messier wrote: "What caused me to undertake the catalog was the nebula I discovered above the southern horn of Taurus on September 12, 1758, whilst observing the comet of that year. This nebula had such a resemblance to a comet in its form and brightness that I endeavored to

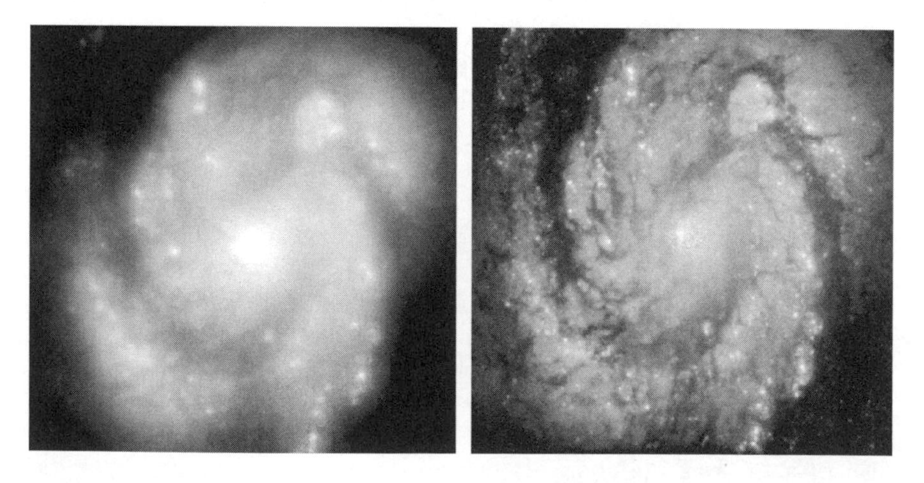

FIGURE 6.2. M100 Galaxy as Seen by the Hubble Space Telescope (2 views)

find others, so that astronomers would no more confuse these same nebulae with comets." Messier irritated many later astronomers by dedicating the comet of 1769 to French emperor Napoleon Bonaparte and interpreting it as an astrological sign of his birth.

By the early 1900s, observational astronomy was in full swing. Hundreds of thousands of celestial objects had been observed. Thanks to the generosity of wealthy patrons and the tireless efforts of a group of women astronomers (see the section titled "Bigger Telescopes and Bigger Stellar Distances"), locations, brightness, and some spectral characteristics of celestial objects had been appropriately cataloged. (See Idea Folder 14, Compiling Star Catalogs.) But distances were known for only the few hundred closest stars, and the detailed nature of nebulae and their distances were unknown. The observers were way ahead. The theorists, however, were about to make substantial progress.

Einstein's Cosmological Input

A major contribution to astronomy's theoretical understanding of galaxies originated in Switzerland. Marcel Grossmann was a member of a class that was graduated in 1900 from Eidgenössische Technische Hochschule (ETH), the Swiss Federal Institute of Technology in Zurich, Switzerland, prepared to teach mathematics and physics.

One of Grossmann's friends disliked school and the especially harsh academic regimen of the time but survived because Grossmann shared his notes before exams. Grossmann and two others in the group received appointments as assistants at ETH, but the classmate who disliked school was unable to get anything more than temporary teaching positions. He wrote to Grossmann in 1901: "I have given up the ambition to get to a university." Finally, Grossmann's father recommended this fellow to the director of the patent office in Bern, and, in 1902, the man got a job as technical expert third class in the Bern patent office. During the next seven years, while working as a patent examiner, Grossmann's friend became very productive, published several scientific papers, and earned a doctorate from the University of Zurich. His dissertation, titled "On a new determination of molecular dimensions," was dedicated to Grossmann. Figure 6.3 is a photo of, from left to right, Marcel Grossmann, his friend, Gustav Geissler, and Marcel's brother Gerald, taken around 1900:

Marcel Grossmann's friend and classmate at ETH was none other than Albert Einstein. Although Grossmann became a renowned mathematician, his fame stopped far short of his friend. Nevertheless, before much time had elapsed, Einstein needed Grossmann's help again.

FIGURE 6.3. Marcel Grossmann, Albert Einstein, Gustav Giessler, and Gerald Grossmann

Einstein found the work at the patent office interesting, but his personal interests ranged wider. With his friends philosopher Maurice Solovine and mathematician Conrad Habicht, Einstein formed a group they called the Olympia Academy. Their wide-ranging discussions were of great value to Einstein. An even larger influence on Einstein, however, was exerted by Michele Angelo Besso. Einstein got Besso a job in the patent office, and they walked to or from work together every day for several years. Einstein called Besso "the best sounding board in Europe" for scientific ideas—and Einstein had plenty of them.

The year 1905 was one historians refer to as Einstein's "Annus Mirabilis," year of miracles. That year the respected *Annalen der Physik*

(*Annals of physics*) published five papers by him, covering such topics as the photoelectric effect, a new way of determining the size of molecules, Brownian motion, special relativity, and the equivalence of mass and energy. (For details, see Idea Folder 15, Einstein's Works: Relativity Plus.) In the relativity paper, Einstein unified Newtonian mechanics with James Clerk Maxwell's electricity and magnetism, and explored the consequences of replacing the concept of absolute time and space with an absolute speed of light.

Two years later, Einstein considered how Newton's gravitation would have to be modified to fit into his relativity ideas. In what he termed "the happiest thought of my life," he came up with the idea of the complete equivalence of a gravitational field and the corresponding acceleration of a reference frame. In simplest terms, this principle says an observer in a rocket ship couldn't tell the difference between acceleration of the rocket ship and the effects of gravity, based on any measurements within the rocket ship. This synthesis, called the Equivalence Principle, was the beginning of general relativity.

The next few years brought several life changes to Einstein. In 1912, he was appointed to the faculty at ETH. Scientifically, he became aware of a giant difficulty with his theory of relativity, namely that if all accelerated reference frames are equivalent, then Euclidean geometry cannot hold true in all of them. Einstein remembered studying differential geometry (geometrical relationships among infinitesimally small quantities) in school, but details were lacking.

Fortunately, one of Einstein's colleagues at ETH was none other than Marcel Grossmann, who was by now a well-established mathematics professor. Grossmann helped Einstein learn about differential geometry and tensor calculus, a mathematical study involving multidimensional variables. Einstein wrote, "in all my life I have not labored nearly so hard, and I have become imbued with great respect for mathematics, the subtler part of which I had in my simple-mindedness regarded as pure luxury until now." Einstein and Grossmann wrote a paper jointly in 1913 that almost completely described the final theory of general relativity. The paper, "Outline of a Generalized Theory of Relativity and of a Theory of Gravitation," contained a set of field equations for gravity, but these were not yet in their final form.

Over the next two years, Einstein published papers, consulted with colleagues, wrote other papers, consulted more, published more, until his final general relativity paper was published on November 25, 1915. In December 1915, he said of himself: "That fellow Einstein suits his convenience. Every year he retracts what he wrote the year before." When applied to the orbit of the planet Mercury, Einstein's equations predicted a slight advance of the point of closest approach to the Sun (perihelion), which could not be explained by Newtonian gravitation. Since Mercury's orbit behaves in exactly this fashion, Einstein's theory corresponded nicely to reality and, as a result, attracted much interest from his scientific colleagues.

When general relativity principles were applied to the universe as a whole, some of Einstein's colleagues (especially Dutch astronomer Willem de Sitter) pointed out that his theory implied that the universe as a whole was not stable in a static condition. According to the equations, the universe would have to be either expanding or contracting. Given the state of astronomical knowledge in 1917, Einstein assumed that

TO AN OBSERVER APPROACHING THE SPEED OF LIGHT, EINSTEIN AND HIS SURROUNDINGS APPEAR TO BE TALL AND THIN

there were no special places, directions, or boundaries to the universe and that the universe as a whole was motionless. To his chagrin, he found that, in order to keep the universe stationary, he needed to add a term to his equations to counteract the attractive effect of gravity. The term he added was called the cosmological constant. Although several astronomers tried to talk him out of it, he remained steadfast about the need to include this constant.

Bigger Telescopes and Bigger Stellar Distances

By 1920, astronomy was poised to solve two giant problems: the size of the Milky Way and the nature of nebulae. Major contributors to these efforts were George Ellery Hale (see Figure 6.4) and Henrictta Swan Leavitt (see Figure 6.5) born within a week of each other in 1868.

George Ellery Hale was born into a Chicago family of great wealth. In his youth, Hale began a lifelong observational career with a used four-inch refracting telescope. As an undergraduate in physics at the Massachusetts Institute of Technology, he invented an instrument called a spectroheliograph for studying solar prominences and, in 1890, submitted its design as his undergraduate thesis. Hale's application of physics to astronomy invented the field of astrophysics.

Hale's major contribution, however, was in telescopes—big telescopes. Although it might be tempting to think that a telescope's primary function is to magnify, merely making a dim or fuzzy image larger does nothing to improve it. In fact, the major functions of a telescope are to collect as much light as possible and to resolve details. The larger the telescope, the more light it collects, and the greater its ability to distinguish between two closely spaced sources of light. Hale applied his great organizational and money-raising skills to the task of building the largest telescope in the world. He was successful—three times over. Hale-built telescopes held the world's record for size three times in a row. The second telescope Hale built was called the Hooker in honor of Los Angeles businessman John D. Hooker, who donated the money to buy the mirror. This 100-inch- (2.5-meter-) diameter telescope was situated on Mount Wilson, above Los Angeles, and began operation in 1918.

Not only did Hale raise the money and oversee the construction of the Hooker, he staffed the Mount Wilson Observatory with keen analysts, including two freshly-minted Ph.D.'s: Harlow Shapley (who graduated from Princeton in 1914) and Edwin Hubble (University of Chicago, 1917); more about them shortly. In 1928, George E. Hale retired from the Mount Wilson Observatory, citing overwork and a desire to return to personal research. That retirement was short-lived. Soon he began planning

Figure 6.4. George Ellery Hale

and financing another huge telescope, the 200-inch (5-meter) one located on Mount Palomar in California. Hale died in 1938. The 200-inch telescope was completed 10 years later and named in his honor. (For almost 40 years, the Hale telescope held the distinction of being the world's largest.)

Henrietta Leavitt was also born in 1868, also interested in science in her youth, and also went to college in Massachusetts. She was graduated in 1892 from Radcliffe College, then called the Society for the Collegiate Instruction of Women. In her senior year, Leavitt became interested in astronomy, and took another course after graduation. A serious illness caused her to become deaf, but her interest in astronomy continued. In 1895, she became one of Edward C. Pickering's "computers" at Harvard College Observatory. This group of women performed calculations and analyzed photographic data in an attempt to "accumulate the facts," in Pickering's words. At first, Leavitt worked as a volunteer, but after devel-

FIGURE 6.5. Henrietta Swan Leavitt

oping her skills at determining star brightness from photographic plates for seven years, she was appointed to the full-time staff (at 30 cents per hour—equivalent to $6 an hour in 2003 money).

"Pickering's harem," as the women were affectionately called, were not independent researchers. They did as they were directed. Leavitt's tedious but significant task was to catalog particularly unusual stars found in the Small Magellanic Cloud, a fuzzy patch of starlight that resembles a broken-off piece of the Milky Way. Both the Small and Large Magellanic Clouds are quite apparent to observers in the Southern Hemisphere. They are named after Ferdinand Magellan, who recorded seeing them on his voyage around the world in 1519. Within the Magellanic Clouds, Leavitt discovered 1,777 stars whose brightness, or luminosity, varied in a regular fashion from bright to dim and back.

Stars whose luminosity vary regularly are called Cepheid variables, because the first one was found in the constellation Cepheus. Cepheid variable periods are quite regular, ranging from 1 to 100 days. By painstaking comparisons of photographs taken at different times, Leavitt found a correlation: *Brighter stars had longer periods.* Using this correlation, called the period-luminosity relation, a Cepheid variable's brightness and period could be used to find its distance from Earth. Leavitt did not, however, apply this potentially valuable distance-measuring technique to significant astronomical questions. Instead, Leavitt, who, according to a colleague, "possess[ed] the best mind at the Observatory," published the

Period-Luminosity relation in 1912 and moved on to her next assigned task. She continued to work at the Harvard College Observatory until she died in 1921. Her death was regarded as a "near calamity" by her colleagues.

The great Danish astronomer Ejnar Hertzsprung recognized the value of Leavitt's period-luminosity relationship and, in 1913, used it to determine the distance to the Small Magellanic Cloud. Harlow Shapley also used the technique at Mount Wilson to find the distances to another set of star groupings called globular clusters. When globular cluster locations were plotted three-dimensionally, it turned out that the center of the distribution was 15,000 parsecs (later revised downward to 9,000 parsecs) from our solar system. Shapley concluded that the center of the globular cluster distribution was also the center of the Milky Way galaxy. He thus estimated the overall size of the Milky Way as 100,000 parsecs, considerably larger than all previous estimates. Further, Shapley's study of a nova (a newly observed star) in the Andromeda nebula allowed him to estimate its distance as 10,000 parsecs. Thus, Shapley's picture of the universe was that it was one big galaxy, the Milky Way, and the solar system is located far from its center.

One Big Galaxy versus Many Separate Galaxies

The contrast between Shapley's model of the Milky Way and the more conventional one came into sharp focus in 1920 at a meeting of the National Academy of Science in Washington, D.C. The young Harlow Shapley (see Figure 6.6) was invited to give the William Ellery Hale (George's father) lecture that year. But rather than a straight presentation, the lecture was set up as a debate. Shapley was joined by the Lick Observatory's Heber D. Curtis (see Figure 6.7), who had just completed a survey of spiral nebulae.

The topic of their debate was "The Scale of the Universe." Curtis argued the standard view of the time: The Milky Way was only about 10,000 parsecs in diameter, and Earth is near its center. In his concluding remarks, Curtis departed from the confines of the stated topic and ventured that spiral nebulae (as they were called at the time) are quite distant and constitute separate galaxies. (We now know they are spiral galaxies.) Although Shapley was not prepared for this issue, he believed spiral nebulae were small gas clouds within our galaxy, and cited as evidence recent observations of a Mount Wilson colleague and personal friend, Adrien van Maanen. Curtis dismissed van Maanen's work as being unsubstantiated. In fact, van Maanen's observations later turned out to be faulty.

Figure 6.6. Harlow Shapley

While the debate had no clear winner and wasn't even well attended, Shapley's idea of a larger Milky Way with Earth far from its center seemed to catch the public's attention. Back at Mount Wilson, Shapley's colleague Edwin Hubble made no secret of his sympathies for Curtis. Hubble and Shapley had never gotten along since Shapley had worked on a project Hubble wanted while Hubble went off to France to fight in World War I. Also, Hubble's British mannerisms, a constant reminder of his Rhodes Scholar studies abroad, irritated Shapley. Their competition was cut short in 1921. Edward C. Pickering had died, and Harlow Shapley left Mount Wilson to head the Harvard College Observatory in 1921. Hubble turned his attention to M31, which he thought he might be able to resolve into individual stars and perhaps even determine its distance from Earth. The 100-inch Hooker telescope beckoned.

FIGURE 6.7. Heber Curtis

A Universe of Galaxies

After many long, cold nights at the telescope, Hubble, shown at work in Figure 6.8, was rewarded. On the evening of October 5–6, 1923, he found the first Cepheid variable star in M31. Using Henrietta Leavitt's Cepheid Period-Luminosity relation and the telescope built by George Hale, Hubble determined the distance to M31 (now known as the Andromeda galaxy) to be 300,000 parsecs. Even using Shapley's inflated diameter of the Milky Way, M31 was definitely too distant to be located within our galaxy.

The "island universe" idea was now backed by evidence. It turned out that there were two different kinds of Cepheid variables, so Hubble's distances were actually too small. The modern estimate of the distance to M31, the Andromeda galaxy, is even farther away, 750,000 parsecs.

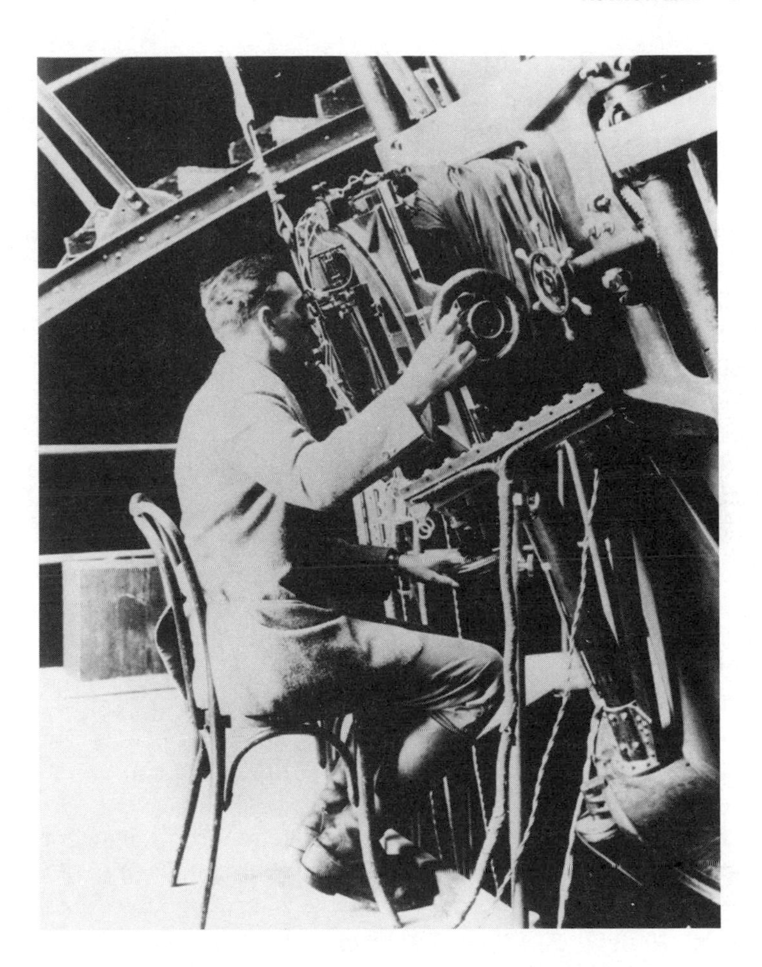

FIGURE 6.8. Edwin Hubble

Thanks to Hubble, astronomy's picture of the universe had to change significantly. The Milky Way had become only one of many galaxies, scattered over huge expanses of space.

Next, Hubble turned his attention and the Hooker telescope to the task of determining the detailed structure of galaxies. For several years, Hubble observed the fuzzy patches of light that had been such a nuisance to Messier's comet search. He found many were actually galaxies of stars. Finding spirals, barred spirals, elliptical, and irregular galaxies, he categorized them in terms of their shapes. Hubble published this scheme in a form that was called the "tuning-fork diagram" because of its shape. By 1929, the importance of Hubble's contribution to astronomy had been assured with his evidence for the distant location of galaxies and their categorization. Hubble, however, had even greater work ahead.

Determining the Speeds of Galaxies

Understanding Hubble's crowning achievement requires a look at a familiar phenomenon in a different context. Think of yourself driving along the highway, minding your own business. Suddenly you hear a sound behind you and look into your rearview mirror. Sure enough, it's a police car, siren wailing.

You glance at your speedometer. Your speed's within the legal limit now, but what about a mile or so ago, when you passed that car? Much to your relief, the police cruiser speeds by. But you notice an odd thing. The sound of the siren was higher pitched when the police car was bearing down on you than when it was going away.

This isn't your imagination, it's a real phenomenon, called the Doppler effect. When a sound wave is emitted by a moving source, the frequency heard by a stationary observer is different from the frequency emitted: If the source approaches the receiver, the sound is higher pitched; if the source moves away from the receiver, the sound is lower pitched. You hear this same high-pitch, low-pitch pattern as a train goes by, or a race car, or an airplane. The faster the sound source moves, the more noticeable the frequency shift.

The Doppler effect also works for light. If a source of light approaches an observer, the light is shifted toward a higher-frequency end of the spectrum, referred to as a blueshift; if the source is receding, the light is shifted to a lower frequency, called a redshift. Since our experience doesn't include extremely fast speeds, the Doppler effect for light is not noticeable. But scientists using spectroscopic instruments to measure the amount of frequency shift can calculate the speed of the source of light. On Earth, weather forecasters use Doppler radar to obtain the speed of frontal systems; police officers use it find out how fast you are driving. Applied to astronomy, the Doppler effect allows the determination of the speed of stars or even whole galaxies.

The first astronomer to use the Doppler shift was Vesto M. Slipher, who worked at the Lowell Observatory in Flagstaff, Arizona, for his whole career, from 1901 to 1952. In 1912, besides searching Mars for canals, which was wealthy amateur astronomer Percival Lowell's pet project, Slipher began measuring the Doppler shifts of spiral nebulae, even before they were recognized as galaxies. The first one he measured, M31, had the huge speed of 300 kilometers per second (km/s), the largest speed ever measured. It was blueshifted—meaning it was coming toward us. By 1917, Slipher had measured the speeds of 15 spiral nebulae and found that 13 of them were redshifted, indicating motion away from us, and in some cases at speeds far greater than the 300 km/s of M31. The implications of this mad rush away from the solar system were not fully appreciated at the time, although there must have been some speculation about possible reasons for our system's apparent unpopularity.

Here's where Hubble came in again. For determining the speeds of galaxies, Hubble relied on Doppler shifts found by Slipher and by Hubble's colleague Milton Humason, who eventually measured the recession speed of 800 galaxies. Humason started at Mount Wilson as a mule driver, then worked his way up to night watchman, assistant astronomer, and eventually observer and coauthor with Hubble on important papers. Not bad for someone with a fourth-grade education.

Hubble set about finding the distances to the galaxies for which Slipher and Humason had found speeds. Henrietta Leavitt's Cepheid analysis was accurate for the nearest ones, but not for the galaxies that were farther away. Cepheids in distant galaxies were just too faint. Hubble devised a new distance-measuring method based on the brightest star in a galaxy. The "brightest stars" technique allowed distance estimates for all but the last few galaxies on Slipher's list. For those, Hubble used the total amount of light from the galaxy as the basis for his distance estimation.

Observing Universal Expansion

To see what type of relationship there was between distance and velocity, Hubble plotted a graph showing velocity as a function of distance (see Figure 6.9). In spite of the scatter of the measurements, it was clear that the relationship was a linear one. That is, *the farther away a galaxy is, the faster it moves.* Strictly speaking, this relationship applied only to the galaxies Hubble chose. However, when generalized, it implied something remarkable and unexpected: *The universe as a whole is expanding.*

To see how this happens, consider a simpler analogy. Suppose there is a race, the Cosmic Marathon. When the race begins, some runners take off at 4 miles per hour, some at 3 miles per hour, and others at 2 miles per hour. One hour into the race, the 4-mph group would have covered 4 miles, the 3-mph group 3 miles, and the 2-mph group 2 miles, producing a graph just like the one generated by Hubble. Note that from *any* runner's perspective, it seems that all other runners are moving away.

The linear relationship between galaxy recession speed and distance is now called Hubble's law in his honor. Although the distances he determined have been corrected by modern measurements, Hubble's

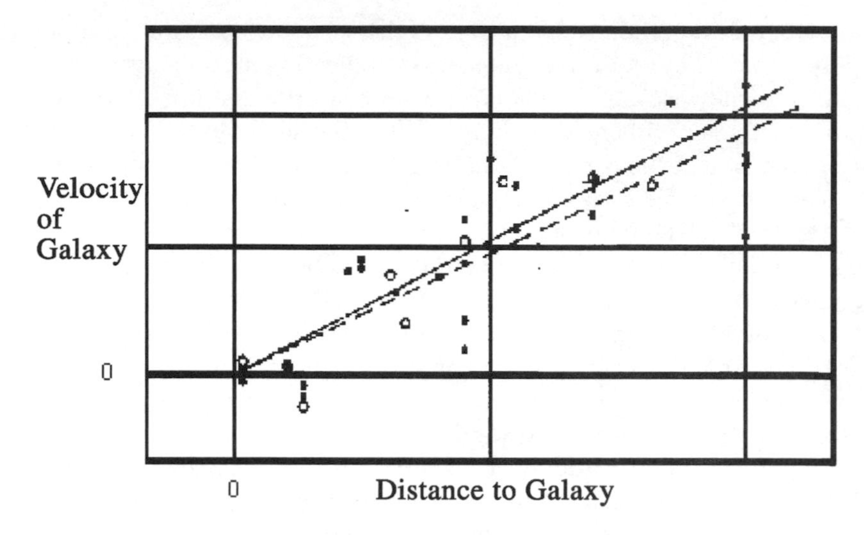

FIGURE 6.9. Hubble's Original Graph (Redrawn) of Velocity as a Function of Distance

fundamental conclusions remain valid. The universe consists of galaxies of stars and is huge and expanding. Figure 6.10 shows Edwin Hubble, second from left, and Albert Einstein, third from right.

When Einstein learned of Hubble's work, he abandoned the cosmological constant he had added to his equations of general relativity to make the universe static. In retracting this term, Einstein called it "The biggest blunder of my life." As we'll see, the cosmological constant may be making a comeback as a possible solution to the biggest unsolved problem in astronomy.

Discovering Dark Matter

It didn't take long for theoreticians to realize that if the expanding universe of galaxies was extrapolated into the past, at an earlier time all the matter and energy of the universe must have been very close together. The theory that resulted was derisively named the "Big Bang" by Fred Hoyle, the champion of a rival theory. The theory and the name stuck, however, because of supporting experimental evidence. (See Idea Folder 16, The Big Bang.)

Remarkably, a huge discrepancy regarding the masses of galaxies was discovered shortly after the presentation of Hubble's law and Einstein's cosmological constant retraction, but this inconsistency was ignored for almost 40 years. Even more amazingly, the astronomer who

FIGURE 6.10. Mount Wilson Library in 1931. *From the left:* Milton Humason, Edwin Hubble, solar astronomer Charles St. John, Albert Michelson, Albert Einstein, University of California president William Campbell, and Walter Adams, director of the Mount Wilson Observatory. Behind them is a portrait of Mount Wilson founder George Hale.

first noticed the problem had graduated from ETH in Switzerland like Einstein and had spent his professional career at California Institute of Technology (Caltech), Mount Wilson, and Mount Palomar like Hubble.

His name was Fritz Zwicky. Born in Bulgaria in 1898, Zwicky went to Switzerland to live with his grandparents at age six and remained a Swiss citizen all his life. Too young for World War I, Zwicky studied theoretical physics at ETH, where he applied quantum mechanics to crystals for his Ph.D. thesis in 1922. In 1925, Zwicky moved to the United States on a Rockefeller fellowship, selecting to study at Caltech because its location in Pasadena's foothills bore some small resemblance to his beloved Alps. Although his sponsor, Robert A. Millikan, expected Zwicky to focus on quantum mechanics, he became attracted to astronomy. He began collaborating with another German-speaking astronomer, Walter

Baade. Early in his career, Zwicky studied the cluster of galaxies known as the Coma Berenices cluster, listed by Messier as M100.

Using the Doppler techniques pioneered by Vesto Slipher and carried out at Mount Wilson by Milton Humason, Zwicky found the velocities of eight of the galaxies in the Coma cluster and estimated the mass needed to keep these galaxies gravitationally bound to the cluster. Next, he compared that mass to a calculation of the cluster's mass based on the light it gives off. It turned out that a lot more mass was needed to keep the cluster from flying apart. Zwicky called this missing mass "*dunkle materie*—dark matter." His calculations implied that there had to be much more dark matter than ordinary matter in the Coma cluster. This alarming result was largely ignored by other astrophysicists for almost 40 years—perhaps because it was published in German *"Die Rotverscheibung von Extragalaktischen Nebeln"* (The Redshift of Extragalactic Nebulae) in a little-read journal, *Helvetica Physica Acta.*

During a long and fruitful career, Zwicky held a wide assortment of ideas of checkered quality, all pursued with relentless conviction. Some thought him brilliant, others regarded him as belligerent. Almost everyone who met Fritz Zwicky, who is shown in Figure 6.11, had an opinion about him. Perhaps the way he often greeted visitors to Caltech, "Who the devil are you?" should be applied to dark matter. Whatever the reason, Zwicky's dark matter didn't make a big impact on astronomy for a while.

The next major contribution was made in 1970 by Vera Rubin and W. K. Ford, who first studied the rotation of M31 (the Andromeda galaxy) and then more than 60 other spiral galaxies. It turned out that these galaxies were all rotating faster than their visible mass could support, again implying the existence of unseen mass. As more experimental evidence came in, the problem became too big to ignore. Dark matter does seem to exist, and there is almost 10 times as much of it as ordinary bright (visible) matter—unless we revise our ideas about gravity (more later).

In the Dark about Dark Matter

Three distinct ways of explaining the nature of dark matter are under consideration: baryonic dark matter, nonbaryonic dark matter, or a possible misunderstanding of gravity.

BARYONIC DARK MATTER Although, strictly speaking, only protons and neutrons are baryons (see chapter 2), astronomers include electrons in baryonic dark matter. The point is, this kind of dark matter consists of well-known particles but doesn't emit enough radiation to be detected.

FIGURE 6.11. Fritz Zwicky

Possible examples of baryonic dark matter include:

- *Ordinary matter.* Clouds of helium and hydrogen scattered throughout the intergalactic medium qualify as ordinary dark matter. This is also called dim matter.
- *MACHOs. MA*ssive *C*ompact *H*alo *O*bjects consist of objects in the outer reaches of galaxies (the halo) that have mass but, because of their small size or minimal radiation, are undetectable to us. Examples of MACHOs include:
  - Brown dwarfs, which are objects intermediate in size between Jupiter and the smallest star, approximately 80 times Jupiter's mass. These objects would have been formed at the same time as stars and planets, but since they had insufficient mass

to begin nuclear fusion, they simply cool slowly by radiating energy too dim for our sensors to pick up.

- White dwarfs, neutron stars, and black holes, which are the remnants of previously existing stars of small, medium, and large mass that give off too little radiation (or none, in the black hole's case) to be detected.

Current efforts to find MACHOs include gravitational lensing, in which light from distant stars is bent by the presence of a MACHO, revealing its presence indirectly. Results of experiments in the Milky Way suggest there are few MACHOs in the outer regions of the galaxy's halo—not enough to account for very much dark matter.

NONBARYONIC DARK MATTER Nonbaryonic dark matter would be made of particles that are not among the known set of fundamental particles that have mass. There may be both cold and hot nonbaryonic dark matter.

- *Cold dark matter.* This matter would consist of extremely massive, slow-moving particles. These particles have been called WIMPs—*W*eakly *I*nteracting *M*assive *P*articles. None has been found experimentally, but some have been postulated as part of theories of how fundamental particles got their mass (see chapter 2). Cold dark matter might consist of:
    - Photinos, which are supersymmetric partners of photons with mass 10 to 100 times that of a proton.
    - Axions, which are hypothetical particles postulated to explain the lack of a particular property of the neutron and also to account for observed asymmetries in the universe.
    - Quark nuggets, which are unusual, not-yet-observed combinations of the six quarks (see chapter 2).
- *Hot dark matter.* This matter would consist of light-mass particles that move very fast. The neutrino is the most likely candidate for hot dark matter. Neutrinos were originally thought to have zero mass, but recent experiments imply that they may have a small mass. Although there may be many neutrinos in the universe, their aggregate mass is likely so small that they do not affect the dark matter problem appreciably.

MISUNDERSTANDING OF GRAVITY Galaxies are still modeled as if they are a collection of particles that obey Newton's laws. Although gravitational theory has withstood all experimental tests so far, new experiments may reveal the necessity of modifications at intergalactic distances.

Predicting the Future of the Universe

While dark matter is clearly a serious and unsolved problem, it is *not* the biggest problem facing astronomy today. That problem developed in the late 1990s, as cosmologists were studying the overall development of the universe from a theoretical perspective. By looking at the growth of the universe in graphical terms, several distinct possibilities for the universe's overall development in space and time are noted (see Figure 6.12).

A simple way of analogizing the motion of the universe is to throw a ball into the air here on Earth. If you throw it pretty fast, it will rise high into the air, eventually stop momentarily, then come back down toward your hand. This situation would occur in a closed universe. The reason the ball returns is that the force of gravity caused by Earth's mass is sufficient to pull it back. Now throw the same ball the same way while standing on a small asteroid. If the asteroid is tiny enough, the ball might be going fast enough to escape the asteroid's gravitational attraction and never return. This situation would occur in an open universe. If you stand on a body of exactly the right mass, the ball will arrive at infinite distance with zero velocity, the situation that occurs in a critical universe.

So, in a sense, the question about the overall development of the universe seemed to boil down to this: Is the mass of the universe

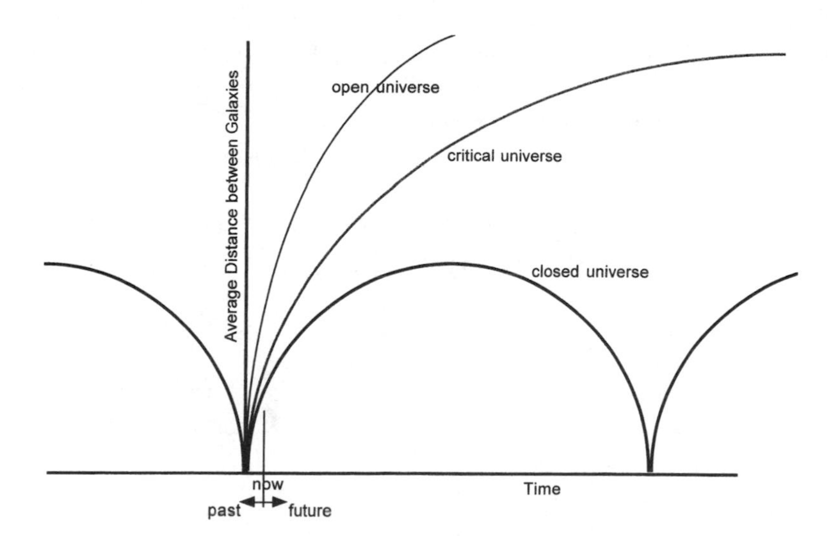

FIGURE 6.12. Distance between Galaxies in the Universe as a Function of Time

sufficient to halt the universe's expansion or not? This question would make the chief determinant of the universe's development its overall density of matter and energy. Both mass and energy must be included since they are interconvertible according to the famous Einstein relation $E = mc^2$. (See chapter 2.)

The matter/energy density is usually expressed as a ratio, called Omega, of the matter/energy density relative to a particular value (the critical density). An Omega of exactly 1, where the matter/energy density is exactly the value of the critical density, implies expansion of the universe at a slightly decreasing rate, toward infinite distance in infinite time, where it would neither expand nor contract further. This case is referred to as a critically dense universe. If the presence of mass determines the geometry of space-time, the critical density corresponds to the flat universe. In a flat universe, parallel lines remain parallel and Euclidian geometry holds.

An Omega greater than 1 means the universe's expansion would decelerate faster, the universe's size would reach a limit, then it would reverse direction and eventually recollapse, resulting in what is called the Big Crunch. This case is called the closed universe. Lines that start out parallel would converge in this non-Euclidean case.

An Omega less than 1 implies that the universe would expand forever, with its rate of expansion decreasing only slightly in time. This case is the open universe. Parallel lines would diverge in this non-Euclidean geometry.

Based on visible matter/energy, Omega is substantially less than 1, which implies an open universe. Current estimates of the amount of dark matter in the universe add considerably more mass, but the total still doesn't approach critical density. On the basis of visible and dark matter, the universe is open, regardless of whatever the details of dark matter may turn out to be. Case closed? Not by a long shot.

Observing the Unexpected: Acceleration of the Universe

In the early 1990s two separate groups of scientists set out to measure distant supernovae (see chapter 3), expecting to determine the deceleration of the universe by finding its expansion rate at an earlier time and comparing it to the present rate of expansion, which they presumed to be slower than it was in the past. What they found wasn't deceleration at all. Their results showed just the opposite: *acceleration*. The scientists were so surprised that, to make sure there were no errors, they analyzed the results several times before releasing them.

Before we examine these data, let's set the scene for what they were trying to accomplish. Recall that Hubble's difficulty in finding the distance to faraway galaxies was based on the fact that Cepheid variable stars in distant galaxies were so dim. It would seem reasonable that a better way would be to seek brighter objects with known luminosity and then determine their distances based on their relative luminosity. Although supernovae are extremely bright, their luminosity depends on their mass. However, one particular kind of supernova involves a star of fixed mass and hence known luminosity. This one occurs when a white dwarf receives mass from a companion star, and the mass is just enough to push it past the white dwarf mass limit (1.4 times the mass of the Sun).

The white dwarf then explodes completely in what is called a Class Ia supernova. Because of their extreme luminosities, Class Ia supernovae are easily spotted in distant galaxies. Such supernovae always explode with the same absolute luminosity, so their distance may be determined by measuring their apparent luminosity: The dimmer the supernova, the more distant it is. One difficulty with this approach is that Class Ia supernovae remain near peak brightness for only a few weeks.

In 1998, the Supernova Cosmology Project at Caltech's Lawrence Berkeley National Laboratory and the High-z Supernova Team (an international consortium) analyzed different Type Ia supernovae near peak brightness and determined their distances. Using the Doppler shift technique pioneered by Vesto Slipher, they determined the redshifts of the galaxies in which the supernovae were located and compared them with the Hubble relationship. The measurements showed that these distant supernovae were substantially dimmer than the Hubble relationship predicted. Since light from these events has taken 4 to 8 billion years to reach us, the measurements indicate that the universe is expanding more rapidly now than it was in the past. In other words, the rate of expansion of the universe is *accelerating*.

The following year, a more distant supernova was found. It turned out to be the most distant one observed so far, with its light originating 11 billion years ago. This supernova is brighter than expected. So, 11 billion years ago, the early universe's expansion must have been decelerating because of gravity. But then, 4 to 8 billion years ago, the universe started to accelerate and galaxies began to spread apart at an ever-faster rate.

The strong implication of this measurement is that whatever is causing the acceleration of the universe now was less important or even absent during the universe's early stages. It became important about halfway through the universe's expansion and has since become dominant. The situation is like a driver who slows down while approaching a

red traffic light, then stomps on the accelerator as soon as the light turns green.

In the Dark about Dark Energy

What is this "stuff" that is causing this cosmological speedup? We don't really know, but we've already given it a name. The missing mass/energy has never been seen, so it's dark. And since it acts in opposition to gravity, it can't have mass in the usual sense. University of Chicago astrophysicist Michael Turner dubbed it *dark energy* in 1999.

Thanks to a set of experiments from a completely different perspective, we have an estimate of how much of this unknown dark energy there is, even though we don't know *what* it is. Several different experiments were designed to probe the overall geometric properties of space to determine if the universe is open, flat, or closed. The background microwave radiation that fills the entire universe is left over from the original Big Bang. During the first 400,000 years after the Big Bang, the early universe was still so hot that it was opaque to electromagnetic radiation. Then it cooled enough, and radiation was emitted. During those 400,000 years, the radiation could travel only a limited distance, so any fluctuations in it would be limited in size. However, in traveling since then, the fluctuations in radiation would be distorted by the overall curvature of space. Measuring the size of the minute temperature fluctuations within this radiation allows us to determinate the overall curvature of space. High-altitude balloons and a sensor atop the weather station at the South Pole have been used to measure these fluctuations. Projects BOOMERANG, MAXIMA, and DASI analyzed these fluctuations and determined that the overall geometry of the universe is flat: Omega is 1 to within ± 4% (see Figure 6.13).

A flat universe has an Omega of 1, so the mass/energy density must be exactly the critical value. Since ordinary matter and dark matter together make up about 27% of the critical mass/energy density, in order to make the overall geometry of the universe flat, the remaining 73% must be dark energy. This theory leaves us in the odd position of being able to estimate the amount of dark energy out there but being completely in the dark about its nature.

Here is the picture these data present: After an initial burst of inflation, the universe settled down to an expansion whose rate was decreased

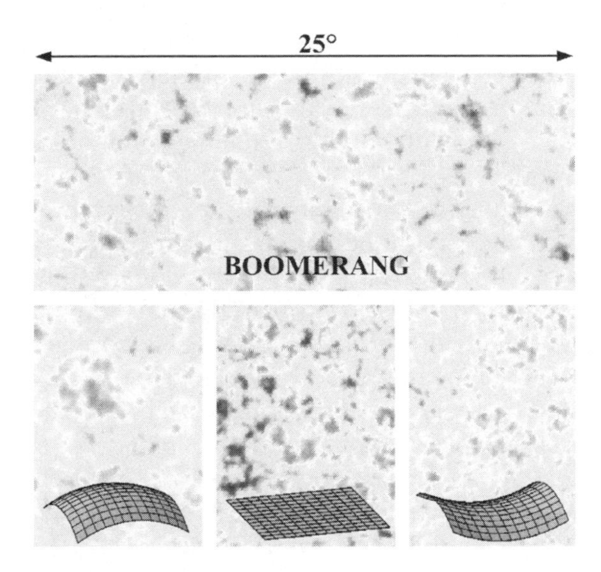

Figure 6.13. Background Microwave Fluctuations Determining the Overall Curvature of Space. The top panel displays experimental data; the bottom panels are predictions of the three possibilities, with two-dimensional analogs of the space-time curvature. From left to right, the cases are Closed, Flat, and Open. The data clearly matches the Flat Universe prediction most closely.

by the presence of matter (ordinary and dark). Dark energy was unimportant in the early stages and must have been so smoothly distributed that it didn't interfere with the formation of galaxies and clusters. Then a few billion years later, dark energy began to take over and exert negative pressure, opposing gravity and causing the universe to accelerate. At the present time, dark energy is slightly stronger than gravity, but as the universe expands faster, the increased distances will weaken gravity's influence further. Dark energy will become even more dominant in the long term, leading to increased acceleration of the universe's expansion.

Solving the Puzzle: Where, When, and How, and Who?

From a theoretical standpoint, there are several possible ways to explain dark energy:

- *Reintroduce Einstein's cosmological constant.* It would be ironic if Einstein's "biggest blunder" turned out to be a necessity after all. Yet a cosmological constant of just the right size would supply the negative pressure that would allow the universe's expansion to accelerate, in agreement with experimental results. But if the cosmological constant represents the zero-point energy inherent in the vacuum (a theoretical idea from quantum mechanics, related to the Heisenberg Uncertainty Principle), it is 120 orders of magnitude too high and must somehow be adjusted downward.
- *Add a time-varying term to Einstein's field equations.* If some quantity in Einstein's equations varied with time, that could possibly explain the relative unimportance of dark energy in the early universe and its more-recent dominance. Although theoreticians would prefer fundamentally simple equations with as few adjustable parameters as possible, this less-elegant possibility must be considered.
- *Consider a time variation in fundamental quantities previously assumed to be constant.* Perhaps the speed of light or the gravitational constant changed in value over time. Research is being done on these possibilities, but results are slow in coming and controversial.
- *Add a fifth, as-yet undiscovered interaction.* This interaction is termed quintessence, and it is represented by an as-yet undetected field of negative energy that permeates all of space. A related idea is a spinning field, referred to as spintessence.
- *Consider hypothetical particles called axions.* If axions exist, photons may oscillate into axions and then back into photons, causing variations in stellar luminosity. Another possibility is that axions may

connect dark matter and dark energy in some way. Axions are exotic particles that may be related to the problem of how particles in the universe got their masses (see chapter 2).

- *Consider the possibility of multiple universes.* Perhaps the quantum foam gave rise to many universes, and we are in one of a multitude. Other universes might have different force strengths, different constants, or possibly altogether different physical laws. Ours harbors life, and that's how we are able to speculate about the nature of the universe.
- *Speculate on collisions between the membrane containing our universe and membranes containing other universes.* If one of the theories about the source of particle masses (see chapter 2) that involve multiple unsensed dimensions turns out to be correct, perhaps the membrane on which our universe exists interacts with other membranes through gravity. The membranes may then collide, which would cause us to revise all previous theories about universe development.

WHERE, WHEN, AND HOW Besides theoretical efforts, several experiments are planned to clarify the nature and extent of dark energy and dark matter.

- *James Webb Space Telescope.* The Hubble Space Telescope is scheduled to be replaced in 2010 by another space-based telescope with substantially enhanced capabilities. Considering the success of the Hubble Space Telescope, its replacement should produce a wealth of data.
- *Planck Satellite.* The European Space Agency plans to launch a satellite to carry out more precise measurements of the fluctuations in the cosmic microwave background than are currently available. This project is scheduled for launch in early 2007.
- *The Sloan Digital Sky Survey.* This ambitious project, already under way, uses a dedicated 2.5-meter telescope to map the location of galaxies in one-quarter of the entire sky. More than 100 million galaxies will be included.
- *SNAP.* The SuperNova/Acceleration Probe project plans to launch one space-based telescope capable of measuring 2,000 Class Ia supernovae per year for the project's three-year lifetime. Although still in the proposal stage, this project could begin operation as early as 2006.
- *The 2df Galaxy Redshift Survey.* This is a project of the Anglo-Australian Telescope at Siding Spring in New South Wales, Australia. The redshifts and spectra of more than 250,000 galaxies will be surveyed using a 3.9-meter telescope. This survey is currently in progress, and periodic updates are issued on the Web at *www.aao.gov.au/2df/*.

WHO In March 2000, the National Academy of Sciences formed a Committee for the Physics of the Universe. Charged with the task of assessing the interdisciplinary area of science between the fields of physics and astronomy and providing a broad vision extending beyond traditional categories, this group addressed opportunities to explore new science relevant to both fields. Their report stressed "deep connections . . . between quarks and the cosmos" and recommended an "interagency initiative on the physics of the universe with the participation of DOE [Department of Energy], NASA, and NSF [National Science Foundation]."

Members of the Committee for the Physics of the Universe and other people likely to make progress toward a solution are:

Michael Turner, University of Chicago (chair)
Roger D. Blandford, Caltech
Sandra M. Faber, University of California, Santa Cruz
Thomas K. Gaisser, University of Delaware
Fiona Harrison, Caltech
John P. Huchra, Harvard University
Helen R. Quinn, Stanford Linear Accelerator Center
R. G. Hamish Robertson, University of Washington
Bernard Sadoulet, University of California, Berkeley
Frank J. Sciulli, Columbia University
David N. Spergel, Princeton University
J. Anthony Tyson, Lucent Technologies
Frank A. Wilczek, MIT
Clifford Will, Washington University
Bruce D. Winstein, University of Chicago
P. J. E. Peebles
John Bahcall
Jeremy Ostriker
E. W. "Rocky" Kolb

The universe is like a present someone has brought to a party. The present is quite dark and wrapped in dark paper, but it is festooned with bright ribbons of riotous colors, patterns, and shapes.

So far, we've been so engrossed with the universe's dazzling ribbons of visible matter that we still don't know much about the dark matter and dark energy inside.

We're just now beginning to shake the box.

How will it rattle?

Problem Folders

> Now my own suspicion is that the universe is not only queerer than we suppose, but queerer than we can suppose.
>
> —*J. B. S. Haldane*

Limiting the number of unsolved problems in science is like demanding the mighty Mississippi flow through a garden hose. Although we have explored our five biggest unsolved problems in some depth, a host of other problems claim the attention and effort of scientists. Some of these may rival or even ultimately overshadow our five.

The Problem Folders list and discuss briefly some of the other unsolved problems in science. More information about these and other problems is available from several sources. Check Resources for Digging Deeper at the end of the book.

Physics Problems

What is the nature of light?

Light behaves like a wave in some experimental situations and like a particle in many others. So, which is it? Neither. Particles and waves are simply models that approximate the behavior of light. Light is actually not either particles or waves. Light is something more sophisticated than these oversimplified models can portray.

What are the conditions inside black holes?

Black holes, discussed in chapters 1 and 6, are usually still-collapsing cores of large stars that have undergone a supernova explosion. They are so dense that not even light travels fast enough to escape them. Because of the highly compressed nature of black holes, the usual laws of physics don't apply to

"WE WORK WITH PARTICLES, IMAGINARY NUMBERS, CONCEPTS... ANYTHING WHICH IS DIFFICULT TO GRASP."

their interiors. Furthermore, because nothing can escape from black holes, no experiments can be performed to test any theories about their internal workings.

How many dimensions are there in the universe, and can we devise a Theory of Everything?

As discussed in chapter 2, theories in physics that attempt to go beyond the Standard Model may eventually clarify the number of dimensions as well as provide a "Theory of Everything." Don't let the name of the theory mislead you, however. While a Theory of Everything would provide an excellent base for understanding fundamental particles, as the sheer number of items on the list of unsolved problems indicates, such a theory would still leave many interesting and important questions unresolved. Like Mark Twain's death, the end of science resulting from a Theory of Everything is greatly exaggerated.

Is time travel possible?

In a theoretical sense, time travel is allowed by Einstein's theory of general relativity. However, the necessary manipulation of black holes and their theoretical cousins, worm holes, would require huge amounts of energy, substantially beyond that available through our current technological know-how. For a lucid description of time travel, see *Hyperspace* by Michio Kaku, Oxford Press, 1994, and *Visions* by Michio Kaku, Anchor Books, 1997, or the Web site *http://mkaku.org.*

Will gravitational waves be detected?

Several observatories are searching for evidence of the existence of gravitational waves. If such waves are found, these vibrations in the very fabric of space-time will signal violent cataclysms in the universe, such as supernovae, black hole collisions, and possibly events not now foreseen. For more details, see "Ripples in Spacetime" by W. Wayt Gibbs.

What is the lifetime of a proton?

Several of the theories beyond the Standard Model (see chapter 2) predict the decay of protons, and a number of detectors have been designed to detect that decay. Although no protons have been observed to decay yet, a lower limit for the proton half-life has been established as about 10^{32} years (1 followed by 32 zeros—far longer than the universe has existed). As more sensitive detectors become available, proton decays may be observed or the lower limit for the proton half-life will be raised.

Are high-temperature superconductors possible?

Superconductivity occurs when the electrical resistance of a material drops to zero. Under this condition, an electric current established in the material continues to flow without the energy loss that accompanies normal current flow in materials such as copper wires. The phenomenon of superconductivity was first demonstrated at extremely low temperatures (just above absolute zero, −273 °C or −459 °F). In 1986, researchers were able to make materials superconduct above the boiling temperature of liquid nitrogen (−196 °C or −320 °F), a commercially available commodity. While the fundamental mechanism for this phenomenon is incompletely understood, research continues, with the ultimate goal of achieving superconductivity at room temperatures, which will reduce electrical energy losses.

Chemistry Problems

How do the components of a molecule determine its shape?

Knowledge of the orbital structures of the atoms in simple molecules allows us to determine fairly easily the shape of the molecule. However, for complex molecules, especially biologically significant ones, theoretical analyses of their shapes are lacking. One aspect of this problem is protein folding, discussed in Idea Folder 8.

What is the chemistry of cancer?

Biological factors such as heredity and environmental influences may play large roles in cancers. If the chemical reactions involved in cancerous cells were known, perhaps molecules could be designed to interrupt these reactions and make the cells cancer-resistant.

How do molecules accomplish communications in living cells?

The signaling process in cells involves molecules of the right shape "fitting" together in complementary ways, carrying a message. Protein molecules are most significant, and their folding patterns influence their shape. Thus, increased knowledge of protein folding may help solve this communication problem.

What is the molecular basis of cell aging?

Another biochemical problem, aging, may involve DNA and various proteins that repair DNA that has been shortened by repeated replications. (See Idea Folder 9, Genetic Technologies.)

Biology Problems

How does an entire organism develop from a single fertilized egg?

This question is one that probably will be answered once the major problem in chapter 4 is solved: What is the complete structure and function of the proteome? Certainly, variations in protein structure and function will occur from one organism to another, but substantial commonalities undoubtedly will be found.

What causes mass extinctions?

Five mass extinctions of species have occurred in the past 500 million years. Their causes continue to be the subject of research. The last extinction, which happened 65 million years ago at the boundary between the Cretaceous and Tertiary eras, was the one that rendered dinosaurs extinct. As David Raup's book *Extinction: Bad Genes or Bad Luck?* (see Resources for Digging Deeper) puts it, were the majority of life-forms that existed at the time wiped out by genetic factors or by cataclysm? A hypothesis put forward by the father and son team of Luis and Walter Alvarez said that a large meteorite (about 10 km or 6 miles in diameter) crashed into Earth 65 million years ago. This impact created vast dust clouds that prevented photosynthesis for a long enough time to kill many plants as well as species of animals all the way up the food chain to the large and vulnerable dinosaurs. Supporting evidence for this hypothesis came from a large meteorite crater found in the southern Gulf of Mexico in 1993. Is it possible that prior extinctions were the result of similar impacts? Research and much debate continues.

Were dinosaurs warm-blooded or cold-blooded?

British anatomy professor Richard Owen named dinosaurs (meaning "terrible lizards") in 1841, when incomplete skeletons of only three species had been found. The process of re-creating the extinct animals fell to British wildlife painter and sculptor Benjamin Waterhouse Hawkins. Since the first species discovered had teeth like an iguana, his models looked like giant iguanas, and created a sensation when exhibited publicly. Since lizards are cold-blooded, the initial assumption about dinosaurs was that they were too. Later several scientists hypothesized that at least some dinosaurs were warm-blooded. Evidence for this hypothesis was lacking until 2000, when a fossilized dinosaur heart was found in South Dakota. Apparently four-chambered, this heart supports the warm-blooded hypothesis, since lizard hearts have only three chambers. More evidence is needed, however, to convince others that this hypothesis is correct.

What is the basis for human consciousness?

Although the human sciences dimensions of this problem take it far beyond the scope of this book, many of our scientific colleagues have attempted to study this intriguing problem.

There are several strong schools of thought about human consciousness, as you might expect. One is a reductionist point of view that says the brain is just a vast set of interacting molecules and that we will eventually unravel the rules for their operation. (See Crick and Koch's article "The Problem of Consciousness.")

Another idea has its roots in quantum mechanics. It says we cannot understand the nonlinearity and unpredictability of brain function until our understanding of the linkage between atomic and large-scale behavior of matter improves. (See *The Emperor's New Mind* by Roger Penrose.)

An older perspective says that the human mind has mystical components that are not amenable to scientific reasoning, so science is incapable of ever understanding human consciousness.

In view of the recent efforts of Steven Wolfram to generate organized patterns by the repeated application of simple rules (see chapter 5), it won't be surprising to see this approach applied to human consciousness and adding still another point of view.

Geology Problems

What causes large Earth climate changes, such as global warming and repeated ice ages?

Ice ages have been common over the past 35 million years, appearing every 100,000 years or so. Glaciers have advanced and retreated across the north temperate zone, leaving their calling cards in the form of rivers, lakes, and moraines. Over 30 million years earlier, during the time dinosaurs roamed

Earth, the climate was much warmer than today, even allowing trees to grow near the North Pole. As the discussion in chapter 5 outlined, Earth's surface temperature depends on the balance between energy arriving and leaving. Many factors influence this balance, including the Sun's energy output, any space debris through which Earth travels, cosmic ray influx, variations in Earth's orbit, atmospheric variations, and changes in the amount of energy Earth emits (albedo).

This is an active area of research, especially in view of the greenhouse gas emission debate that has raged in recent times. Theories are numerous, but real understanding still eludes us.

Can we predict volcanic eruptions or earthquakes?

Some volcanic eruptions can be forecast, such as the recent one (1991) at Mount Pinatubo in the Philippines, but others are not amenable to current techniques and still catch volcanologists unaware (think Mount St. Helens eruption in Washington on May 18, 1980). Many variables come into play in the eruption of volcanoes. No single unified theoretical understanding applies to all volcanoes.

Earthquakes are even harder to predict than volcano eruptions. Some prominent geologists have expressed doubts that reliable prediction is even possible. (See Idea Folder 13, Earthquake Prediction.)

What's happening in Earth's cores?

The innermost two layers of Earth, the outer and inner cores, are inaccessible because their great depth and pressure prohibit direct measurement. Whatever geologists have learned about the cores comes from inferences about surface and overall density, composition and magnetic properties, plus seismic wave studies. Additionally, iron meteorites are studied because of the similarity between their formation and that of the overall Earth. Recent seismic wave results have revealed differing wave speeds in north-south waves in comparison to east-west waves, which implies a layered solid inner core.

Are we alone in the universe?

While there is no experimental evidence for extraterrestrial life, there is no shortage of theories about such beings and efforts to detect signals from distant civilizations. (See Idea Folder 4, Extraterrestrial Life.)

How do galaxies evolve?

As you'll recall from the discussion in chapter 6, Edwin Hubble categorized all known galaxies on the basis of their shapes. Although nicely descriptive of present states, this scheme doesn't lend itself to understanding how galaxies evolve. Several theories have been proposed to explain how spiral, elliptical, and irregular galaxies are formed. These theories are based on the physics of the gas clouds that preceded the galaxies. Supercomputer modeling studies have provided some insights but have not yet pointed to a uniform theory of galaxy formation. Development of such a theory requires more study and data.

Are Earth-like planets common?

Mathematical models have predicted anywhere from very few Earth-like planets to billions of them within the Milky Way galaxy. Precision telescopes have detected more than 70 planets outside the solar system, but most are the size of Jupiter or larger. As telescopic searches become more sophisticated, additional planets will likely be found and will help determine which mathematical model is most realistic.

What is the source of gamma-ray bursts?

Just about once a day, a burst of gamma rays lights up the sky. This energy burst is often brighter than all other gamma-ray sources put together. (Gamma rays are similar to visible light but have a much higher frequency and energy.) This phenomenon was first observed in the late 1960s but not published until the 1970s because the detectors were being used to monitor possible Nuclear Test Ban Treaty violations.

At first, astronomers thought the sources of these bursts were probably within our Milky Way galaxy. The bursts' high intensity implied they must be close.

Yet as more data were gathered, it appeared that the bursts come from every direction rather than being concentrated in the plane of the Milky Way galaxy.

A 1997 sighting tracked by the Hubble Space Telescope was identified as emanating from a region offset from the center of a faint galaxy located several billion light-years away. Since the source is not near the galaxy's center, it isn't likely to be powered by black holes. Rather, the current thinking is that these gamma-ray bursts come from the normal stars contained in the galaxy's disk, possibly resulting from something like colliding neutron stars or other objects with which we are unfamiliar.

Why is Pluto so different from all the other planets?

The four inner planets—Mercury, Venus, Earth, and Mars—are relatively small, rocky, and close to the Sun. The four outer planets—Jupiter, Saturn, Uranus, and Neptune—are large, gaseous, and far from the Sun. Then there's Pluto. It has an icy surface and an orbit unlike any other planet. Pluto is small like the inner planets and far from the Sun like the outer planets. In this sense, Pluto is a misfit. It orbits the Sun near a region called the Kuiper Belt, which contains many objects similar to Pluto. (Some astronomers refer to these objects as "Plutinos.")

Recently a few museums have decided to demote Pluto from planet status. Until many other Kuiper Belt objects are mapped, controversy about Pluto's status will continue.

How old is the universe?

The age of the universe may be estimated by several techniques. One technique estimates the age of the elements in the Milky Way, based on radioactive decays of elements with known half-lives, and then assumes the elements are made (in large-star supernovae) at a constant rate. According to this method, the universe's age would be 14.5 ± 3 billion years.

Another technique involves estimating the ages of star clusters, based on assumptions about cluster dynamics and distances. The oldest clusters are estimated to be 11.5 ± 1.3 billion years old, making the universe about 11 to 14 billion years old.

Using the universe's expansion rate and the distance to the farthest objects to estimate its age yields about 13 to 14 billion years. Recent findings about the increased expansion rate (see chapter 6) have made this estimate more uncertain.

Most recently, a completely independent method was developed. The Hubble Space Telescope, working near its observational limits, has measured the temperature of the oldest white dwarfs in the globular cluster M4. The technique used is similar to estimating how long ago a fire burned by measuring the temperature of its embers. By this method, the oldest white dwarfs turn out to be 12 to 13 billion years old. Assuming that the first stars formed less than 1 billion years after the Big Bang, this yields a 13- to 14-billion-year estimate of the overall age of the universe and provides a check for the other methods.

In February 2003, data returned from the *Wilkinson Microwave Anisotropy Probe (WMAP)* spacecraft allowed the most accurate determination of the universe's age to date: 13.7 ± 0.2 billion years.

Are there multiple universes?

One of the possible solutions to the accelerating universe problem explored in chapter 6 involves multiple universes on separate "branes" (multidimensional membranes). Although this idea is purely theoretical, it provides fascinating ammunition for speculation. For more detail about multiple universes, see the book *Our Cosmic Habitat* by Martin Rees and the article by Dennis Overbye: "A New View of Our Universe."

When will Earth next be hit by an asteroid?

Earth is constantly being hit by space debris. Vital questions include how big are the objects that hit us and how often do they hit. Objects 1 meter (1 yard) in diameter enter Earth's atmosphere several times a month. They often explode at high altitude, releasing energy equivalent to a small atomic bomb. About once a century, a 100-meter (100-yard) object arrives, making a bigger impression (pun intended). About 700 square miles of trees were flattened when such an object exploded over the Tunguska forest in Siberia in 1908.

The impact of a 1-km object every million years or so can create widespread damage and even climatic changes. A 10-km object's impact is probably what killed off the dinosaurs in the K-T (Cretaceous-Tertiary) extinction 65 million years ago. Although an object of this size may arrive only once every 100 million years, we have taken steps to avoid being caught unaware. Near-Earth Objects (NEOs) and Near Earth Asteroid Tracking (NEAT) projects have been undertaken, with the goal of identifying and tracking 90% of the estimated 500 to 1,000 asteroids greater than 1 km in diameter by 2010. Another program, Spacewatch, at the University of Arizona, scans the skies for potential Earth-impactors.

For more information, see the Web sites: *http://neat.jpl.nasa.gov, http://neo.jpl.nasa.gov,* and *http://spacewatch.lpl.arizona.edu/.*

What happened before the Big Bang?

Since time and space began with the Big Bang, the concept of "before" has no real meaning. It's almost like asking what is north of the North Pole. Or, as American author Gertrude Stein might have put it, there wasn't any "then" then. Such difficulties, however, never stop theorists. Perhaps before the Big Bang, time was imaginary; perhaps there was nothing at all, and the universe came from a fluctuation in the vacuum; or perhaps there was a collision with another "brane." (See the multiple universes question discussed earlier.) These theories will have a difficult time finding experimental support, since the primeval fireball's enormous temperatures disassembled any atomic or subatomic structures that might have existed prior to the expansion.

Idea Folders

Many of the ideas contained within this book are discussed only briefly as they relate to the biggest unsolved problems in science. Yet readers may be interested in more detailed descriptions of items that are mentioned only in passing. The Idea Folders allow you to delve into subjects in more depth. They are arranged in the order of appearance within the chapters, and each contains resources if you wish further information. Additional information is contained in the "Resources for Digging Deeper" at the end of the book.

We hope these ideas help satisfy your curiosity—or even pique it further. In the future, solutions to some of these problems will be found and raise many more problems in the process.

1 Anti-matter

Almost every particle has a corresponding anti-particle. Generally, anti-particles have the same mass as their standard partner and charges of equal strength but opposite sign. As you can see from Figure I.1, every quark has a corresponding anti-quark, (anti-up, anti-charm, . . .), every lepton has a corresponding anti-lepton (anti-electron neutrino, anti-muon neutrino, . . .), and W^+ and W^- bosons are particle anti-particle counterparts. Only the photon, the Z boson, the gluon (there are eight of these), and the hypothetical graviton have no anti-particle. Alternatively, they are said to be their own anti-particles.

As discussed in chapter 2, anti-matter originated as a theoretical prediction, made when British physicist P. A. M. Dirac integrated quantum mechanics with special relativity in 1928. A similar but simpler mathematical analogy would be the solutions of the equation $x^2 = 9$, which are $x = +3$ and $x = -3$. Often, if an equation has two solutions, one is discarded on the basis of being nonphysical. Scientists must have been tempted to discard the solution of the Dirac equation, which allowed for the existence of a particle similar to an electron but carrying a positive rather than a negative electric charge. Yet within four years, American physicist Carl D. Anderson demonstrated the anti-electron (positron) experimentally in cosmic ray track experiments, so evidence supported the prediction. The anti-proton was observed in 1955 at the University of California, Berkeley by Emilio Segré and Owen Chamberlain, and the anti-neutron showed up a year later.

The event that created the electron and positron in Anderson's cloud chamber in 1932 is referred to as pair production or creation. A photon of light from cosmic rays surrenders all its energy, which is then turned into mass according to the Einstein equation $E = mc^2$. When an electron and positron meet, their mass is converted completely to energy, and two photons of light stream off in opposite directions. This process is called annihilation, and it converts mass into energy, with the amount again given by the Einstein equation.

In a theoretical sense, nothing would prevent anti-protons from teaming up with anti-neutrons to make anti-nuclei and anti-electrons could be persuaded to join these anti-nuclei to form anti-atoms. In fact, nine anti-hydrogen atoms were created in 1995 at the European Laboratory for Particle Physics by a team led by German physicist Walter Oelert. You might wonder if these anti-atoms caused any mischief in the lab. They did not. The overwhelming preponderance of ordinary matter annihilated the nine anti-hydrogen atoms within 40 billionths of a second.

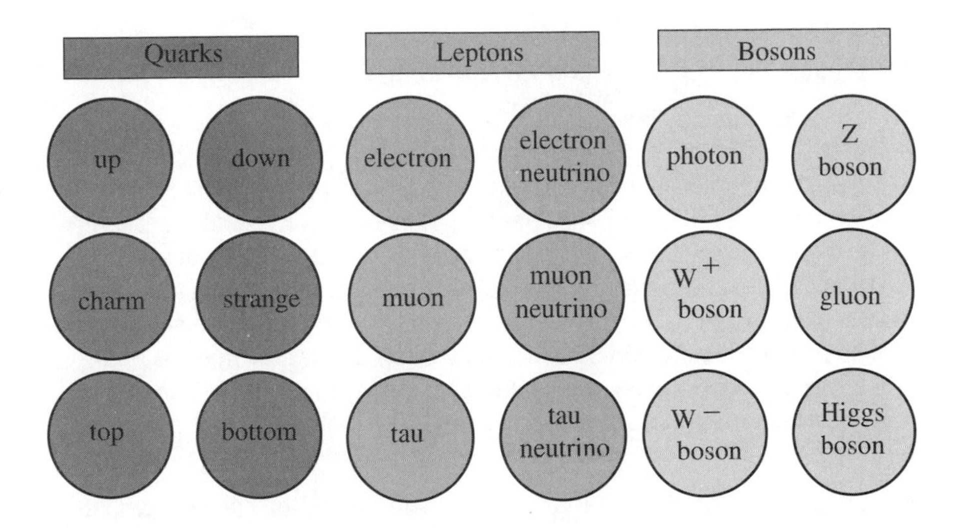

FIGURE I.1. Fundamental Particles

Science fiction uses lots of anti-matter, especially as drives for spaceships. The biggest difficulty with the idea of an anti-matter drive is the containment/contamination problem. While some wonderfully creative technologists are designing the next generation of spaceships based on anti-matter, the safety issue looms large, since 1 gram of matter/anti-matter undergoing annihilation would generate about as much energy as the atomic bombs dropped on Japan in 1945.

On a larger scale, you might wonder if somewhere in a distant galaxy, or even the Milky Way, there might lurk substantial quantities of anti-matter. After all, if our only contact with these galaxies is through the photons of light they give off, we could be fooled. A photon is its own anti-particle, so we couldn't detect the difference between galaxies of matter and galaxies of anti-matter, since both would give off photons. True enough, but the cosmic rays that rain down on us constantly consist of far more than photons—yet they contain no unaccounted-for anti-matter. Further, if proton/anti-proton annihilation were taking place at an anti-galaxy edge, light of a particular frequency would be emitted. No such light is detected. It appears that the universe is almost all conventional matter.

But the absence of anti-matter poses an altogether different problem. If the universe in which we live is symmetrical, then equal amounts of matter and anti-matter should have been created in the Big Bang. If that were the case, they would have annihilated each other completely, and we wouldn't be here to discuss this question. So where did the anti-matter go? One theory says an entire anti-universe was created and is located somewhere else, possibly on some other "brane" from M-theory (see chapter 2).

A recent experimental finding is that there is an asymmetry in the rate of decay of some forms of matter and anti-matter. Mesons, particles made up of two quarks, are not stable and thus are not found in ordinary matter. One kind of meson, the K meson, has been studied extensively. A difference in decay rates for K mesons and anti-K mesons was discovered by physicist C. S. Wu of Columbia University in 1957. Anti-K mesons decay slightly more quickly than K mesons. In 2001, experimenters at Stanford University and in Tsukuba, Japan, announced an asymmetry in the decay of B mesons and anti-B mesons, with anti-B mesons decaying slightly faster. The value of asymmetry will become more accurate as more data are obtained in these long-term studies.

If anti-matter decays slightly faster than ordinary matter, the situation would be similar to the case where an army of 1 million and one fights an anti-army of 1 million. If each soldier kills one enemy, at the end of the battle, there is only one soldier left standing. Matter and anti-matter annihilate each other, but since there was a slightly greater amount of ordinary matter, that's what prevailed. If this is an accurate representation, think of how much matter there must have been before the great annihilation.

Standard Model predictions for symmetry-breaking variations in the decay rates are too small to produce the abundance of matter observed in the universe, but there's always the newer M-theory around the corner.

For more information, see: "Explore: In Search of Antimatter," *Scientific American,* August 20, 2001. Web site:
http://physicsweb.org/article/news/5/3/1/1.

2 Accelerators

As its name implies, an accelerator takes a slowly moving particle and speeds it up. At higher speeds, particles have more energy, so high-energy physics developed concurrently with particle accelerators. The utility of high-energy particles became clear when American physicist Carl D. Anderson discovered the electron's anti-particle partner, the positron, in the tracks left in a cloud chamber after a high-energy collision involving cosmic rays. Since cosmic rays come in all energies, from all directions, and at seemingly random times, a more reliable source of high-energy particles was needed to carry out systematic experiments on fundamental particles.

Linear accelerators take charged particles and accelerate them in an electromagnetic field along a straight-line path, similar to the way electrons are accelerated in electron-gun (not LCD) television sets. A target is set up at the end of the particle's path, and sensors that detect trails left by collision products record the collision aftermath. To reach higher and higher energies, linear accelerators must be longer and longer. The Stanford Linear Accelerator Center has a 2-mile-long (3.2-km) tunnel and accelerates electrons (or positrons) using a standing electromagnetic wave similar to a microwave oven. For more details, see the Web site *www.slac.stanford.edu/.*

The other type of accelerator is the circular accelerator. The first circular accelerator was called the cyclotron, invented by American physicist Ernest O. Lawrence. In 1928, the University of California at Berkeley lured 27-year-old Lawrence from Yale, striving to build a physics department comparable to its already-strong chemistry department. The following year, Lawrence, grandson of Norwegian immigrants, happened to browse through a German electrical engineering journal. He saw sketches of a device proposed by Norwegian engineer Rolf Wideröe to accelerate charges to high energies by running them through an accelerating field twice and switching the field direction so that the charges receive twice the energy. Because of the enormous engineering difficulties involved, Lawrence hesitated initially. However, he hated to lose out on the race to high energies, so, early in 1930, he assigned the task of construction of the apparatus to a graduate student, M. Stanley Livingston. By January 1931, Lawrence and Livingston had a working cyclotron 4.5 inches in diameter, which accelerated hydrogen ions to energies of 80,000 electron volts. (See Figure I.2). In 1939, Lawrence won a Nobel Prize for the invention of the cyclotron. By 1940, 22 cyclotrons were either completed or under construction in the United States, and 11 more overseas.

World War II halted cyclotron development for a time. When it resumed, new wrinkles were added, and the energy increased substantially. The synchrotron was developed, featuring a magnetic field that was adjusted so that as particles accelerated, their paths continued to maintain a constant radius. This allowed a smaller volume where a vacuum needed to be maintained, thus simplifying maintenance of a usable beam.

Next the synchrotron was adjusted so that particles could continue to circulate, with energy added to make up for radiation losses. This was called the storage ring. Finally, two storage rings were built adjacent to each other, and the circulating particles were deflected to crash into each other. This intersecting storage ring configuration produced a great deal of basic information about fundamental particles. In the United States, the largest accelerator is the Fermi National Accelerator Lab (Fermilab), just outside Chicago. Built in 1968, this facility has an accelerator called a Tevatron, the world's highest-energy particle accelerator, which can reach an energy level of 0.980-trillion electron volts (TeV) for each of its particle beams: clockwise-circulating protons and anticlockwise-circulating anti-protons. A proton–anti-proton collision produces an energy of 1.96 TeV at the interaction points.

For more information, see the Web site *www.fnal.gov.*

A huge complex devoted to fundamental research is the European Organization for Nuclear Research (CERN). Located on the border between France and Switzerland, CERN maintains 10 accelerators on its site. Scientists of 80

FIGURE I.2. Ernest O. Lawrence with a Model Cyclotron (about 1932)

nationalities representing 500 universities carry on research at CERN. For more information about CERN, see the Web site: *http://public.web.cern.ch/Public.*

CERN's largest accelerator, the Large Electron Positron Collider (LEP), had a beam path that is 27 km (17 miles) long—the longest in the world. The LEP no longer exists; its tunnel is being refitted for use as a Large Hadron Collider (LHC), in which protons will collide with protons at an energy of 7 TeV. When it begins operation in 2005, it will be the highest-energy machine in the world. For more information about the LHC, see the Web site: *http://lhc-new-homepage.web.cern.ch/lhc-new-homepage/.*

Some theorists have suggested that the new LHC might produce miniature black holes, possibly at the rate of one per second, qualifying it as a "black hole factory." These black holes would decay within a small fraction of a second but in the process might create that much-sought particle, the Higgs boson discussed in chapter 2. It might even happen "in as little as one hour of operation," according to the University of Maryland's Gregory Landsberg, in "Black Holes at the Large Hadron Collider," by S. Dimopoulos, and G. Landsberg, *Physical Review Letters* 87 (2001): 161602.

Web sites: *www.aip.org/history/lawrence/first.htm; www.lbl.gov/Science-Articles/Archive/early-years.html.*

3 Fermions and Bosons

All the particles that make up the universe fall into two fundamentally different categories: fermions and bosons. The roots of this distinction lie in Holland, where Samuel Goudsmit and George Uhlenbeck were graduate students at the University of Leiden in 1925. Goudsmit, who was experimentally oriented, noticed additional splitting in the spectra of light given off by helium atoms. Uhlenbeck, who was more familiar with classical physics, thought the cause of this splitting might be related to some inherent property of the electron. Together they arrived at the notion that the electron possessed some intrinsic angular momentum, or spin.

Quantum mechanics was still in its formative stages at the time, so this idea had the effect of adding a fourth quantum number (the other three were the principal, orbital, and magnetic), called the spin quantum number. Although it is tempting to think of an electron as a tiny, madly spinning top, this picture should not be taken literally. The electron's intrinsic angular momentum, or spin, is $\pm \frac{1}{2}(h/2\pi)$, where h is Planck's constant. Spin is also a convenient way to think about the electron because the spin quantum number has two values $+\frac{1}{2}(h/2\pi)$ and $-\frac{1}{2}(h/2\pi)$—corresponding to spin "up" and spin "down." In 1928, British physicist P. A. M. Dirac's formulation of relativistic quantum mechanics provided the theoretical basis for electron spin; Goudsmit and Uhlenbeck's earlier effort involved extremely fortunate guesswork.

In 1925, Austrian physicist Wolfgang Pauli put forth the idea that no two electrons could occupy the same quantum state at the same location. This Pauli exclusion principle forms the basis for the chemical periodic table of the elements.

When the statistical behavior of groups of electrons was studied, Italian/American physicist Enrico Fermi and Dirac derived the theory called Fermi-Dirac statistics. This analysis was later extended to all other particles with spins that are odd half-integer multiples of $h/2\pi$. These particles, called fermions, include all leptons and quarks. Indeed, the universe's mass consists of fermions.

Particles with zero or integer multiples of $h/2\pi$ were studied separately by Indian physicist Satyendranath Bose in 1924. Working at the University of Dacca, in Bangladesh, Bose sent his analysis to Einstein for comments. Einstein translated the analysis into German and strongly recommended its publication. The following year, Einstein extended Bose's work to encompass all particles other than fermions. The statistical behavior of such particles is referred to as Bose-Einstein statistics. Dirac called particles that obey these statistics *bosons.* All the force carriers—the photon, which carries the electromagnetic force; the gluons, which carry the strong force; and the W and Z particles, which carry the weak force—are bosons.

While no two fermions may exist in the same quantum state, no such limitation applies to bosons. In fact, the more bosons there are in a particular energy state, the more likely other bosons will join them. That fact is the basis of the stimulated emission that occurs in lasers, where photons all line up in phase in the same energy state. This gregarious property also helps explain superfluidity in helium and even superconductivity, when electrons are coupled into pairs and act like bosons. In 1995, a gas of rubidium atoms was cooled to a temperature so low that the gas atoms all existed in the same quantum state. This aggregate is called a Bose-Einstein condensate.

The "loner" nature of fermions and the "joiner" characteristics of bosons make them fundamentally quite different. The difference, however, is critical to the nature of the universe as we know it. For example, if fermions joined each other the way bosons do, an atom's electrons would all congregate in the lowest energy level, and thus there would be no chemistry and no life.

4 Extraterrestrial Life

> I have argued flying saucers with lots of people. I was interested in this: they keep arguing that it is possible. And that's true. It is possible. They do not appreciate that the problem is not to demonstrate whether it's possible or not, but whether it's going on or not.
>
> *–Nobel Prize–winning physicist Richard P. Feynman*

Scientists are as fascinated with the possibility of extraterrestrial life as everyone else. The reality, however, is that in spite of representations in movies, TV programs, books, Web sites, and countless stories in the popular culture, *there is no scientific evidence for life anywhere beyond Earth.* Nevertheless, scientific inquiries are taking place on two fronts, the theoretical and experimental.

Theoretical Inquiries

What kinds of life-forms are possible?

- *Carbon-based life, just like us.* Reflecting the majority opinion, the late chemist Cyril Ponnamperuma, of the University of Maryland, thought that the chemistry of life on Earth might be generalizable to the universe. He said that the data "suggest that the formation and linking of life's building blocks—amino acids and nucleotides—may have been all but inevitable, given the starting chemistry of Earth's 'primordial soup.' " Furthermore, Ponnamperuma said, "If there is life elsewhere in the universe, chemically speaking it would be very similar to what we have on Earth."

Most scientists who offer an opinion agree that despite the toy industry's ubiquitous big-eyed green aliens, any extraterrestrial life-form would look quite different from humans. However, some structural and functional elements might be common. For example, eyelike sensors for detecting photons (perhaps in regions of the spectrum beyond the visible); two eyelike sensors for gauging distances; and a short path to the sensor information processor—a brain—all seem likely. Further, a compact body plan including appendages for manipulation of objects in the environment and some means of locomotion also seem reasonable. In some respects, Hollywood's alien representations might not be too far off.

- *Noncarbon-based life.* Besides carbon, the element most likely to act as a backbone for life is the one directly below it in the periodic table, silicon. When this relationship was first noted in the 1890s, novelist H. G. Wells wrote, "One is startled towards fantastic imaginings by such a suggestion: visions of silicon-aluminum organisms—why not silicon-aluminum men at once?—wandering through an atmosphere of gaseous sulphur, let us say, by the shores of a sea of liquid iron some thousand degrees or so above the temperature of a blast furnace."

Indeed, the chemistry of silicon and carbon is similar in many ways. For example, carbon bonds with four hydrogens to form methane, CH_4, while silicon combines with four hydrogens to form silane, SiH_4. Silicon's chemistry with oxygen is similar in some respects (CO_2 and SiO_2), but a substantial difference occurs. Silicon dioxide forms a three-dimensional lattice, with bonds so strong that SiO_2 is a solid (sand), even at elevated temperatures.

In the biochemistry of carbon-based life, energy is harvested from long chain carbohydrates, which are broken apart using protein enzyme catalysts. The waste products are water and carbon dioxide, both of which are easily disposed of because they are liquid and gas. Silicon-based life would be forced to deal with solid waste, which would present unique disposal problems.

Additionally, carbon-based biologically significant molecules have the property called chirality (see chapter 3), meaning the three-dimensional nature of their bonds causes them to twist toward either the right or the left when they form a helical structure. This property makes for subtleties in metabolism that would be absent in silicon-based life, which has much less capacity for chirality.

Finally, there is the matter of abundance. As of 2002, 113 carbon-based molecules had been discovered in space, as opposed to only 10 silicon-based ones. If silicon-based life-forms do exist, they would seem to occupy a much smaller niche than carbon-based life.

Just how probable is the existence of whole extraterrestrial (ET) civilizations? In November 1961, the National Academy of Sciences convened an informal meeting at Green Bank, West Virginia, to discuss extraterrestrial life. National Radio Astronomy Observatory radio astronomer Frank Drake formulated an equation to focus discussion on the probability of extraterrestrial life on a series of factors, each of which could be estimated separately. The equation, which Drake called the Green Bank equation, became a classic and was renamed the Drake equation:

$$\#\text{of ET civilizations} = (\text{stars/year})(f_{\text{planets}})(f_{\text{life zone}})(f_{\text{life}})(f_{\text{intelligence}})(f_{\text{communicative}})(\text{Lifetime})$$

To use the Drake equation to arrive at an estimate of the number of communicative civilizations (ones that send and receive messages) in the Milky Way galaxy, seven factors must be estimated, where all the *f*'s stand for fractions between 0 and 1.

1. *What is our galaxy's formation rate for stars suitable for generating planets suitable for life?*

Large stars have too short a lifetime and small stars are too cool, so only middle-size stars need be considered here.

2. *What fraction of these appropriately sized stars actually have planets?*

Given our current understanding of planet formation, it would seem most of these stars would have planets orbiting them.

3. *What fraction of planets would orbit their star within a zone where life could form?*

In Earth's case, the presence of liquid water is crucial. Venus is too hot for liquid water, Mars is too cold, so our solar system has just one planet in the life zone—Earth. Further, the role of the Moon might be quite significant. The ebb and flow of tides could have influenced the beginnings of life here by causing pools of water to alternately flood and evaporate, possibly concentrating the "primordial soup" at critical times.

Another unknown in life's development is the role of the massive outer planets, especially Jupiter, in deflecting possible asteroid or comet impactors out of the inner solar system. This action protected Earth from disturbing influences that might have stunted or even stopped the development of life.

4. *On what fraction of appropriately placed planets does life actually arise?*

Estimating this factor usually divides optimists from pessimists. Some, like Nobel Prize–winning Belgian biochemist Christian deDuvé, think that given enough carbon and liquid water, the right temperature, and enough time, life is inevitable. Others cite the myriad complexities in even a single-celled organism and say life is extremely rare, possibly even unique. Scientists differ widely on their estimates of this factor. Some doubt the usefulness of the whole approach because of this wide divergence. After all, without evidence, this is just an educated guess, and shouldn't be taken too seriously.

5. *What fraction of life-forms actually develop intelligence?*

On Earth, many species have shown evidence of intelligent behavior, humans sometimes included. Because intelligence seems to be such a good survival talent, it seems likely that many life-forms would develop it, given enough time.

6. What fraction of intelligent life-forms develop technologies that release detectable signals?

While both humans and dolphins are intelligent life-forms on Earth, only human technologies have generated detectable signals, so numbers like 5% to 50% are typically used for this estimate.

7. For how many years does an intelligent civilization release detectable signals into space?

This estimate provides another vehicle for expressing optimism or pessimism. An optimist might foresee a million-year civilization, whereas a pessimist might look at our own case and proclaim the end is near. Don't forget this equation was originally set up for radio astronomy purposes. A civilization could outgrow radio emissions by developing more efficient alternatives, or let their radios fall into disuse as they moved on to more interesting pastimes. In our case, we have been releasing radio emissions for a little over 100 years, so the earliest transmissions have penetrated space to a distance of 100 light-years.

Multiplying all these factors yields an estimate of the total number of communicative civilizations in the Milky Way galaxy. The numbers range from billions (optimists) to one—us. Drake's original estimate was 10,000. Modern versions often converge around a number of communicative civilizations approximately equal to the number of years a civilization releases detectable signals.

Although some have suggested the Drake equation is a way of encapsulating our ignorance into a small space, it is instructive to think about each of the factors separately. Further, the equation allows another estimate to be made: the average distance between communicative civilizations. For neither pessimistic nor optimistic values of the seven factors above, the average distance between communicative civilizations in the Milky Way galaxy would be hundreds to thousands of light-years. If it takes light several hundred years to travel from one civilization to another, communication would take longer than screechy old modems getting on to the Internet, if you can imagine that. Nevertheless, for a million-year-old, technologically adept, expansionist civilization intent on colonizing the galaxy, a few thousand years of travel to a new world is not unreasonable.

Considering that the solar system has been around for only the last third of the galaxy's existence, many other stars have quite a head start. Might they have developed the necessary technology and set out to colonize the galaxy? Knowing the galaxy's size, and making reasonable assumptions about the speed of their spaceships, it would seem that such a project could be accomplished within a few million years. This is large in terms of individual human lifetimes but small in terms of the galaxy's age. In other words, technologically

"HOLD ON TO YOUR HATS. WE'RE PICKING UP ANOTHER BIG BANG."

advanced civilizations might very well colonize the galaxy, à la *Star Trek, Star Wars,* or other science fiction works.

A group of scientists worked at Los Alamos in 1950 to develop the hydrogen bomb. Their lunchtime conversations often featured free-wheeling questions posed by the great physics genius Enrico Fermi to encourage discussion (mostly by him). Thinking about all the time available for galaxy colonization by extraterrestrials, Fermi remarked, "Don't you ever wonder where everybody is?" This question subsequently has been rephrased to "Where are they?" and is referred to as Fermi's Paradox. Any extraterrestrial theories must deal with this simple but powerful question.

Experimental Inquiries

Regarding the estimates necessary to complete the Drake equation, physicist Philip Morrison said, "That isn't the right question. The question is really, 'Should we do something to find out?' . . . To find out, you must do something empirical."

Science's first experimental effort in this direction was undertaken by none other than Frank Drake. For six hours a day, from April to July 1960, the National Radio Astronomy Observatory's 85-foot dish antenna was set to 1420 MegaHertz (MHz) and pointed at two stars of about the same age as our Sun. Signals from the stars Tau Ceti and Epsilon Eridani produced nothing other than static and one false alarm—a formerly secret military project's signals. Project Ozma, named after the queen of Oz, L. Frank Baum's imaginary land "populated by strange and exotic beings," produced no positive results, but the search for extraterrestrial intelligence was officially under way.

Several other projects have been undertaken to listen for extraterrestrial transmissions and even send information in case "they" are listening. The largest one, called SETI, the Search for Extra Terrestrial Intelligence, started in 1984. For more information, see the Web site *www.seti.org.* The movie *Contact,* based on a novel by astronomer Carl Sagan, portrays many aspects of SETI quite accurately, with Jodie Foster cast in a role that seems quite similar to Jill Tarter, cofounder of SETI. (See *Scientific American* profile, "An Ear to the Stars," November 2002.) Of course, Hollywood's search is more successful than reality.

Other efforts include optical searches using lasers and the SERENDIP project—an acronym for Search for Extraterrestrial Radio Emissions from Nearby Developed Intelligent Populations, supported by science fiction writer Arthur C. Clarke.

Recent books of interest include *Sharing the Universe: Perspectives on Extraterrestrial Life* by Seth Shostak, Berkeley Hills Books, 1998; *Life in Outer Space: The Search for Extraterrestrials* by Kim McDonald, Raintree/Steck-Vaughn, 2000; *Why Aren't Black Holes Black?* by Robert M. Hazen, Anchor, 1997; "Where Are They?" by Ian Crawford, *Scientific American,* July 2000; "Who's Out There?" by Jeff Greenwald, *Discover,* April 1999; *Are We Alone?* by Paul Davies, Basic Books, 1996.

If you want to participate in the SETI project, you can download a program that will obtain data over the Internet and process it on your computer when it would normally be in a screen saver mode, display the signals, and send it back to SETI. To obtain this program, go to *http://setiathome.ssl.berkeley.edu/download.html.*

Here's a far-out possibility to consider: What about dark matter/dark energy–based life? Realizing the lack of interaction between dark matter/dark energy and ordinary matter/energy, it's no wonder we haven't sensed them, or they us. Further, given the preponderance of dark energy/dark matter over ordinary matter, life-forms based on them might be so huge in size or number that we might be a minuscule footnote, totally unknown or ignored by the *real* life-forms of the universe.

5 Amino Acids

An amino acid consists of a carbon (designated the *alpha* carbon atom) with four groups linked to it. (See Figure I.3.) The groups are: a carboxyl group (COO^-), which is an acid; an amino group, (H_3N^+), which is a base; a single hydrogen (H); and a group symbolized by R, a side chain, which varies from one amino acid to another.

When the carboxyl carbon of one amino acid bonds covalently to the nitrogen in the amino part of another amino acid, water is released, and a peptide bond is formed. Protein molecules consist of a long chain of amino acids bonded by peptide bonds.

Within the digestive systems of animals, amino acids are released by the digestion of protein molecules, then carried by the bloodstream to the body's cells, where they are recycled. The amino acids are used to build more proteins according to plans contained in the cell's DNA and carried out by RNA with the assistance of protein catalysts (enzymes). Most of the amino acids

carboxyl end

amino end

COO^-

H_3N^+ — C — H

R

side chain

α carbon

Figure I.3. Amino Acid Molecular Structure

needed by the body can thus be synthesized from already present recycled amino acids. These are called *nonessential* amino acids. Others that must be supplied specifically by diet are referred to as *essential* amino acids.

More than 100 amino acids occur naturally in plants and bacteria, but only 20 are found in animals. The following table lists the names, standard abbreviations, and linear structural formulas of the 20 amino acids found in animals.

Amino Acid	Abbreviation	Linear Structure
Alanine	ala A	CH3-CH(NH2)-COOH
Arginine	arg R	HN = C(NH2)-NH-(CH2)3-CH(NH2)-COOH
Asparagine	asn N	H2N-CO-CH2-CH(NH2)-COOH
Aspartic Acid	asp D	HOOC-CH2-CH(NH2)-COOH
Cysteine	cys C	HS-CH2-CH(NH2)-COOH
Glutamic Acid	glu E	HOOC-(CH2)2-CH(NH2)-COOH
Glutamine	gln Q	H2N-CO-(CH2)2-CH(NH2)-COOH
Glycine	gly G	NH2-CH2-COOH
Histidine	his H	NH-CH = N-CH = C-CH2-CH(NH2)-COOH
Isoleucine	ile I	CH3-CH2-CH(CH3)-CH(NH2)-COOH
Leucine	leu L	(CH3)2-CH-CH2-CH(NH2)-COOH
Lysine	lys K	H2N-(CH2)4-CH(NH2)-COOH
Methionine	met M	CH3-S-(CH2)2-CH(NH2)-COOH
Phenylalanine	phe F	Ph-CH2-CH(NH2)-COOH
Proline	pro P	NH-(CH2)3-CH-COOH
Serine	ser S	HO-CH2-CH(NH2)-COOH
Threonine	thr T	CH3-CH(OH)-CH(NH2)-COOH
Tryptophan	trp W	Ph-NH-CH = C-CH2-CH(NH2)-COOH
Tyrosine	tyr Y	HO-Ph-CH2-CH(NH2)-COOH
Valine	val V	(CH3)2-CH-CH(NH2)-COOH

Source: *http://chemistry.about.com/library/weekly/aa080801a.htm.*

6 Building a DNA Model

DNA's extremely small size makes it impossible to see. This is a source of difficulty for some people because they think of DNA as an abstract, theoretical concept, not an actual molecule. One way to get a better sense of DNA is to build a physical model of it.

Children's construction systems provide an excellent resource for building molecular models, including DNA. One of us (AWW) used the K'NEX system to build a DNA model, shown in Figure I.4 being held by young people who share a great deal of his DNA.

This particular model was made with the K'NEX 32 Model Building Set contained in a Blue Value Tub (34006), available for $30 to $40. (See *www.knex.com.*) Instructions for building a DNA molecule model are contained on the Web site *http://c3.biomath.mssm.edu/knex/dna.models.knex.html.* Although the directions there use only green connectors, they have nice diagrams and show a few other options.

FIGURE I.4. DNA Model Held by Ray, Melissa, and Tim Noe (Some of AWW's Grandchildren)

When finished, you'll have a portion of a DNA molecule that contains 48 base pairs. It will be almost 1 meter (about 3 feet) long.

The finished model is somewhat inaccurate in its simulation of the real DNA. In the model, each blue rod makes about a 20-degree angle with the prior rod, while the hydrogen bonds in the real DNA are parallel to within 6 degrees. However, the model does illustrate several turns of the helix, the major and minor grooves, and the Watson-Crick base pairs of A-T and C-G.

If you build this model, you will be able to visualize the *lac* operon mechanisms, the splitting of the two strands of DNA during reproduction, and the operation of restriction enzymes, which cut the DNA at specific locations because of the way these enzymes "fit" into the molecule.

7 Codons

Nearly all life-forms on Earth use the same genetic code. The key to this code is codons. If the nucleotide bases contained in DNA are thought of as being the letters in the genetic code, then codons are the words, and a gene is a series of codons strung together to make a sentence. As the central dogma of gene expression puts it, DNA's message is transcribed into mRNA (messenger RNA), which is then translated into proteins.

To see how codons work, let's examine this process in detail:

- The sequence of nucleotide bases contained in DNA come in four varieties, adenine, thymine, cytosine, and guanine, commonly represented by A, T, C, and G.
- mRNAs transcribe DNA's nucleotide bases in the same order to the ribosome, except they substitute uracil for thymine. In the ribosome, proteins are built by stringing together amino acids (see Idea Folder 5, Amino Acids) that have been collected. The order of amino acids in the protein is determined by tRNA (transfer RNA), which transfers the order of nucleotide bases originally contained in DNA.

But how do the four nucleotide bases specify which of 20 amino acids must be used to build the protein?

- If each nucleotide base specified a single amino acid, only 4 amino acids could be built.
- If two nucleotide bases in a row specified a single amino acid, $4^2 = 16$ amino acids would result. This is not sufficient.
- If three nucleotide bases in a row specified a single amino acid, $4^3 = 64$ amino acids could be specified, which is more than enough. Thus, a codon needs to consist of a triplet—3 bases in a row.

The triplet nature of codons was confirmed experimentally by Francis Crick in 1961. Determination of which codon triplet of nucleotide bases specifies which amino acid was begun in 1961 by American biochemist Marshall Nirenberg, when he determined that UUU codes for the amino acid phenylalanine. Following more experimental work by Nirenberg and others by 1966 the entire correspondence between codons and amino acids was determined.

The following table lists the correspondence between three-letter codons and the amino acid added to a growing protein molecule by an RNA. The tables use RNA nucleotide bases U, C, A, and G rather than the DNA bases T, C, A, and G. Start and stop codons signal the beginning and the end of RNA transcription.

	U	**C**	**A**	**G**	
U	UUU = Phe	UCU = Ser	UAU = Tyr	UGU = Cys	U
	UUC = Phe	UCC = Ser	UAC = Tyr	UGC = Cys	C
	UUA = Leu	UCA = Ser	UAA = Stop	UGA = Stop	A
	UUG = Leu	UCG = Ser	UAG = Stop	UGG = Trp	G
C	CUU = Leu	CCU = Pro	CAU = His	CGU = Arg	U
	CUC = Leu	CCC = Pro	CAC = His	CGC = Arg	C
	CUA = Leu	CCA = Pro	CCA = Gln	CGA = Arg	A
	CUG = Leu	CCG = Pro	CAG = Gln	CGG = Arg	G
A	AUU = Ile	ACU = Thr	AAU = Asn	AGU = Ser	U
	AUC = Ile	ACC = Thr	AAC = Asn	AGC = Ser	C
	AUA = Ile	ACA = Thr	AAA = Lys	AGA = Arg	A
	AUG = Met	ACG = Thr	AAG = Lys	AGG = Arg	G
G	GUU = Val	GCU = Ala	GAU = Asp	GGU = Gly	U
	GUC = Val	GCC = Ala	GAC = Asp	GCG = Gly	C
	GUA = Val	GCA = Ala	GAA = Glu	GGA = Gly	A
	GUG = Val	GCG = Ala	GAG = Glu	GGG = Gly	G

Note that most amino acids are specified by more than one codon. This redundancy often means that the same amino acid is specified regardless of which base is in the third position in the codon. Since this third position is often the one that is misread, the redundancy minimizes the effects of reading errors.

The following table lists each of the 20 amino acids and the corresponding codons and two protein-building actions.

START	AUG, GUG	Leu	UUA, UUG, CUU, CUC, CUA, CUG
Ala	GCU, GCC, GCA, GCG	Lys	AAA, AAG
Arg	CGU, CGC, CGA, CGG, AGA, AGG	Met	AUG
Asn	AAU, AAC	Phe	UUU, UUC
Asp	GAU, GAC	Pro	CCU, CCC, CCA, CCG
Cys	UGU, UGC	Ser	UCU, UCC, UCA, UCG, AGU, AGC
Gln	CAA, CAG	Thr	ACU, ACC, ACA, ACG
Glu	GAA, GAG	Trp	UGG
Gly	GGU, GGC, GGA, GGG	Tyr	UAU, UAC
His	CAU, CAC	Val	GUU, GUC, GUA, GUG
Ile	AUU, AUC, AUA	STOP	UAG, UGA, UAA

8 Protein Folding

Proteins, the final product of the intricate DNA, RNA, and protein enzyme efforts, accomplish life's heavy lifting—both figuratively and literally. Two fundamentally different kinds of proteins, named globular and structural on the basis of their structure, accomplish quite an impressive array of functions:

- *Enzyme catalysts.* Globular proteins fit snugly around specific molecules and encourage chemical reactions needed for life.
- *Defense.* Different globular proteins guard against particular molecules that "fit" the proteins' shapes.
- *Transport.* Another variety of globular proteins transports small molecules, again based on the shapes of both the protein and the transported molecule. For example, hemoglobin has an oxygen-shaped cavity, transports oxygen in the blood, and releases it when needed. Contrast this with what happens when carbon monoxide enters hemoglobin's cavity and gets "stuck," rendering that hemoglobin unable to transport more oxygen.

- *Structural support.* Collagen is the most abundant structural protein in the body of vertebrates. It is the supporting molecular structure for bones, ligaments, tendons, and skin.
- *Motion.* Molecules of actin and myosin slide, providing muscular contraction.
- *Regulation.* Proteins act as cell surface receptors and internal gene action regulators such as *lac* repressors (see chapter 4).

The shapes of proteins are critical to the accomplishment of their many tasks. These shapes are far from simple. If you think of the long string of amino acids that make up a protein as being like a fiber, then the functional shape of proteins is like a complicated basket woven from that fiber.

The complex, three-dimensional structure of proteins was first noticed in the 1930s, when W. T. Astbury found dramatically different X-ray diffraction patterns in ordinary and stretched human hair. The American chemist Linus Pauling, working with Robert C. Corey in 1951, used his knowledge of chemical bonds to predict that the simpler protein molecules would have structures that looked like helices (called alpha, α) or pleated sheets (called beta, β).

(In England, James Watson and Francis Crick feared that Pauling would determine DNA's structure before they could. It turned out that Pauling was working with faulty data that led him to believe that DNA was a triple helix, rather than the double helix worked out by Watson and Crick in 1953 using Rosalind Franklin's excellent X-ray diffraction data.)

Shortly after Pauling and Corey's predictions, Danish chemical researcher K. Linderstrøm-Lang proposed a hierarchy of protein structure that included four levels, based again on theoretical grounds. (See Figure 3.6.) Current understanding has added two more levels, which we'll discuss after considering some experimental data.

In 1957, chemist John C. Kendrew, after completing a huge project using X-ray crystallography at Cambridge University in England, determined an accurate three-dimensional structure for the protein myoglobin, which transports oxygen in muscles. When the final results were analyzed, Kendrew said, "Perhaps the most remarkable features of the molecule are its complexity and its lack of symmetry." The point is that proteins generally have convoluted, twisted, complicated three-dimensional structures. Even experienced researchers must work very hard to spot regularities in models of proteins. That's why knowledge of protein structural hierarchy is valuable.

The protein's primary structure is determined by the chain of amino acids that RNA builds, according to DNA's specifications. For a protein 100 amino acids long, each position may be held by any of the 20 amino acids, so there are as many as 20^{100} different possible proteins. This huge number (10^{130}), greater than the number of atoms of ordinary matter in the universe, illustrates the abundance of varieties of possible proteins.

The secondary structure is the alpha (α) helix and beta (β) pleated sheet predicted by Pauling. These structures come about because of attraction of positively charged regions of the molecule to negative regions of the same molecule and other electrical effects.

A supersecondary structure, not illustrated, combines two or more secondary structures, called motifs. A fold or crease is commonly a $\beta\alpha\beta$ motif; the so-called Rossmann fold is a $\beta\alpha\beta\alpha\beta$ motif; and another common motif is the β barrel, which is a β sheet joined to form a tube.

A tertiary structure is often formed by reaction of the molecule to water, and packs portions of the molecule into the interior very tightly, so there is very little wasted space inside. This tight packing helps explain why some mutations that substitute an amino acid of different size can alter protein shapes so drastically that they cannot fulfill their normal roles in the organism's metabolism.

A domain, not illustrated, is a portion of a protein, often hundreds of amino acids long, that forms a characteristic shape, independent of the rest of the molecule. Domains would be analogous to knots tied on a long rope.

Quaternary structure is the name given to the situation where two or more amino acid chains, called subgroups, associate to form a single functional protein. For example, hemoglobin consists of two α chain subgroups and two β chain subunits. Sickle cell anemia is caused by a mutation that substitutes a different amino acid on one corner of a β subunit, which creates a "sticky patch" that binds one hemoglobin molecule to another. The resulting molecular chain is too long and doesn't fulfill the normal function.

A protein's primary structure, which is biologically inactive, is also vulnerable to reacting with other molecules, which might alter its structure and functionality. Thus, proteins often go from the primary state to the tertiary or quaternary state within minutes or even small fractions of a second. This process is termed folding. Conversely, when the protein's environmental conditions (temperature, acidity, concentration of ions) change, the protein may change its shape or unfold. This reverse process is called denaturation. As an example, adding salt or vinegar to food denatures the proteins of microorganisms that would normally grow on the food, thus preserving it.

In many cases, proteins can be denatured, then refold themselves back into their biologically active shape and resume functioning as if nothing ever happened. However, sometimes misfolding may occur. For example, if you boil an egg, the proteins in the egg white unfold. However, when the egg cools, they don't refold to re-form the original white. Instead, they make an insoluble mass (think hard-boiled eggs).

Protein foldings and misfoldings are influenced by other proteins, referred to as chaperonins, that normally function to assist folding by increasing its speed and preventing misfolding. More than 17 chaperonins have been identified, some of which even rescue an already misfolded protein by re-folding it. Intense research continues on the misfolding that may lead to Alzheimer's and mad cow diseases.

Because of the huge number of proteins involved and the even larger number of possible foldings they may undergo, research in this area requires supercomputers to consider all the cases. Like the data generated in the SETI project, protein folding calculations can be downloaded onto your home computer and accomplished in a background mode as a screen saver. If you're interested, go to the Web site *http://folding.stanford.edu/.* More than 60,000 computers have run this program, making a significant contribution to the project called Folding@Home.

Additional resource: *www.faseb.org/opar/protfold/protein.html.*

9 Genetic Technologies

Since the operating systems of all living things are based on DNA, the ability to cut DNA, rearrange it, and then put it back together has spawned a whole new set of industries called genetic technologies.

Many plants and animals have already been affected by this technology. For years, animal breeders and agricultural specialists have been modifying DNA by selective breeding. Recently more direct genetic modifications have been undertaken. Herbicide resistance, nitrogen fixation, and insect resistance are a few of the many traits modified. Increased yields of foodstuffs of high nutritional value have benefited human beings.

Direct applications of genetic technologies to humans have ethical implications that must be considered carefully before any implementation is attempted, especially in view of the incomplete understanding of the human proteome and hence the unknown impact of genetic modifications on human traits (see chapter 4).

Indirect applications of genetic technologies are already profoundly affecting humans. The following lists illustrate the range of biotechnology projects currently under way.

Bacteria are used to grow previously difficult-to-obtain proteins valuable to humans. Among these are:

erythropoietin, which stimulates red blood cell production

growth hormone, which enables normal growth to occur

insulin, useful in treating diabetes

interferon, for a variety of diseases still not fully understood

tissue plasminogen, which dissolves blood clots

Human diseases are now being treated directly by gene therapy. Among these diseases are:

AIDS

Alpha-1 antipysin deficiency

Some cancers

Chronic granulomatous disease

Cystic fibrosis

Familial hypercholesterolemia

Fanconi's anemia

Gaucher's disease

Hemophilia

Hunter's syndrome

Peripheral vascular disease

Purine nucleoside phosphorylase deficiency

Rheumatoid arthritis

SCID, severe combined immunodeficiency

Lists like these are incomplete the moment they are published. Diseases are being added daily. For more up-to-the-minute information, consult the following Web sites, which maintain current news about biotechnology:

www.bioethics.net/news/html/biotech.php

http://life.bio.sunysb.edu/biotech/news/

www.mc.maricopa.edu/~tdclark/html/biotechnology_news.html

http://ucbiotech.org/~news/

The development of our understanding of telomeres is an example of how knowledge of genome/proteome functioning can be translated into technology. A repetitive stretch at the end of a chromosome, called a telomere, often consists of the sequence TTAGGG repeated many times. This would be similar to ending a sentence with the words "and so forth, and so forth, and so forth . . ." In a sense, these repetitive sequences might be regarded as "junk" DNA, because they don't code for the building of any protein. Yet each time DNA replicates, one of the repeat sequences physically separates from the DNA molecule, shortening it. After all the repetitive sequences have left, the next time DNA replicates, the bases that drop off are no longer repetitive end caps but part of the protein specification. This action, which relates directly to the aging of a cell, is referred to as the Hayflick limit. The bases that are essential to building a particular protein are no longer present, so the protein doesn't get built properly and thus doesn't fulfill its function. If that protein's role in metabolism is critical to the life of the organism, the result is death.

Suppose the organism uses this protein to fight a particular virus. Earlier in the organism's life, the protein was built properly, and the virus it counteracted was fought off successfully. But once all the repetitive TTAGGG sequences were gone, no more virus-fighting protein was built, so the virus could spread unchecked. Might this be why flaviviruses such as the West Nile variety are more successful against older people?

On the other hand, cancer cells don't age. They reproduce without limit. So what happened to their repetitive TTAGGG sequences that were supposed to drop off? It turns out that there is an enzyme called telomerase that, when active, restores the missing TTAGGG sequences to the chromosome ends, allowing the cell to reproduce well beyond its normal limit. A suggested possible screening strategy against certain cancers would be to look for the presence of activated telomerase. On the other hand, adding telomerase under noncancer-threatening conditions might lengthen an organism's lifetime. Or deactivating telomerase after cancer treatments might cut down on recurrence possibilities. The research that continues on this fascinating subject has wide-ranging pharmaceutical implications.

This explosion in biotechnology was made possible by the mapping of model organisms and human genomes. Yet because the human genome that was mapped was a composite (which included contributions from one of the principal investigators), tailoring pharmaceuticals or gene therapies expressly to individuals is not yet possible.

That situation is about to change. On August 15, 2002, J. Craig Venter announced plans to create a new DNA sequencing center, run by The Institute for Genomic Research (TIGR), the Center of the Advancement of Genomics, and the Institute for Biological Energy Alternatives. The goals of these institutes include sequencing the complete genome of a specific human being, completing the analysis within minutes or hours rather than months or years, and accomplishing this task for $2,000 to $3,000 rather than the hundreds of millions it took for the Human Genome Project. Although Venter says, "Existing technology is not up to any of these tasks," his goal is to make it happen within 10 years. Once this new technology is up and running, Venter plans to sequence all the microbes in a sample of ocean water simultaneously, as a way to monitor ecosystems. While these plans may appear overly optimistic, Venter's track record makes his predictions likely to be borne out.

10 Greenhouse Gases

A greenhouse keeps plants warm because the glass panes allow higher-frequency visible light from the Sun to pass through quite easily, while these same glass panels reflect and thus retain the lower-frequency infrared radiation emitted by the plants inside the greenhouse. The glass panels also trap heated air. As discussed in chapter 5, Venus, Earth, and Mars have surfaces that are substantially warmer because of an atmosphere that functions just like the glass in a greenhouse.

Figure I.5 shows radiation interacting with Earth. Visible light from the Sun is mostly transmitted right through Earth's atmosphere (1 on the figure), with only a small amount reflected by clouds. The energy is partially absorbed by Earth's surface (2) and partially reflected (3). The surface molecules then re-

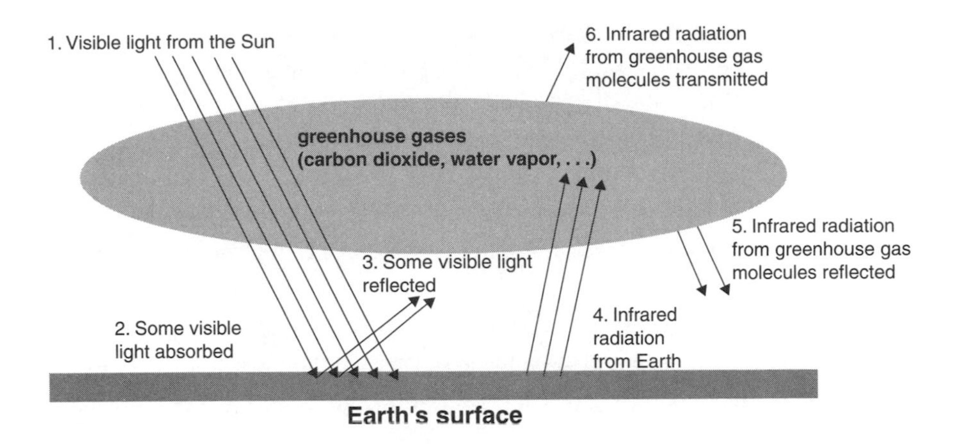

FIGURE I.5. Radiation Interacting with Earth

radiate the energy at a lower frequency, in the infrared region of the spectrum (4). Gases in Earth's atmosphere reflect much of the infrared radiation back to the surface (5), while allowing some of it to return to space (6). The net result is that Earth's surface is heated to a higher temperature, just as a greenhouse is heated.

The vast majority of Earth's atmosphere consists of nitrogen and oxygen, which do not reflect infrared radiation back to the planet's surface. Other atmospheric gases do accomplish this effect and are appropriately called greenhouse gases. Naturally occurring greenhouse gases in the atmosphere include water vapor, carbon dioxide, methane, nitrous oxide, and ozone. Technological applications add substantially to the amount of each of these greenhouse gases and also generate several greenhouse gases that do not occur in nature.

Carbon dioxide makes up 76% of the greenhouse gas total. Naturally occurring sources of carbon dioxide include volcanic eruptions, decay of dead plant and animal matter, evaporation from the oceans, and respiration by oxygen-breathing animals. Carbon dioxide is removed from the atmosphere by absorption by seawater and photosynthesis of both ocean-dwelling plankton and land-dwelling biomass, including forests and grasslands (referred to as "sinks"). Human activities (called anthropogenic activities) that release carbon dioxide into the atmosphere include the burning of solid waste, fossil fuels, wood, and wood products.

Methane, which constitutes 13% of greenhouse gases, is also called swamp gas. Methane is given off by plant decay, especially in rice fields, by bacteria that break down organic matter in wetlands, and in the intestines of many animals (think belching cows). Methane is generated by human activity in the mining and transportation of fossil fuels, the decomposition of solid waste in landfills, and the raising of livestock.

Nitrous oxide makes up 6% of greenhouse gases and is released naturally by the ocean and by bacterial action in soils. Humans add nitrous oxide from nitrogen-based fertilizers, sewage treatment plants, and automobile and truck exhaust.

Roughly 5% of greenhouse gases come from sources that are entirely human made. These include hydrofluorocarbons (HFCs), perfluorocarbons (PFCs), and sulfur hexafluoride (SF_6), which have been used in a variety of industrial processes.

Recent predictions of global warming have evoked critical interest in greenhouse gases. As is often the case with global issues, there are scientific, technological, economic, and ethical issues all bound up together. Most of these are beyond the scope of this book, so let's just explore some of the scientific parts that are related to the discussion of weather in chapter 5.

First, let's look at Figure I.6, which shows past temperature data.

The graph shows that Earth's average temperature has risen by about 1°F in the last 100 years. Additional indications of past global warming include the recession of glaciers, the melting of both north and south polar ice caps, an increase in evaporation and precipitation, and elevation of ocean levels. Clearly, Earth is getting warmer.

But is the cause of this temperature increase the recent increase in greenhouse gases? Let's examine Figure I.7.

The United Nations–sponsored 2,500 scientist Intergovernmental Panel on Climate Change (IPCC) has concluded that greenhouse gases *are* the culprit. (See the Web site: *www.ipcc.ch/*.) You might think that natural sources of greenhouse gases are so much larger than human-generated sources that the increase might be the result of something besides human activities. Actually, atmospheric scientists theorize that natural sources and sinks are approximately in balance, so the increases noted probably result from anthropogenic sources.

Besides adding carbon dioxide through the burning of fossil fuels and wood, deforestation is another human activity that has had a large impact on atmospheric levels. In the tropics, logging and clearing land for farming and animal pastures accounts for 3,500 acres of forest lost per hour. Carbon dioxide is added to the atmosphere as the trees are burned, and deforestation reduces the amount of carbon dioxide sinks available globally.

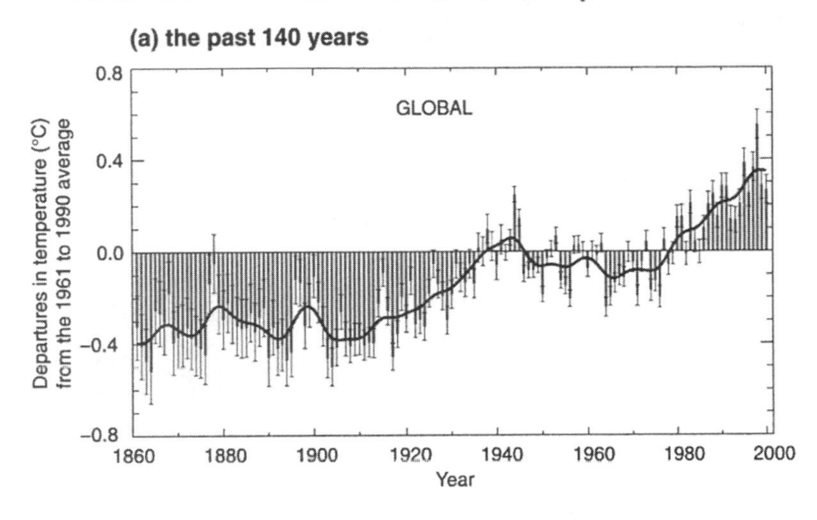

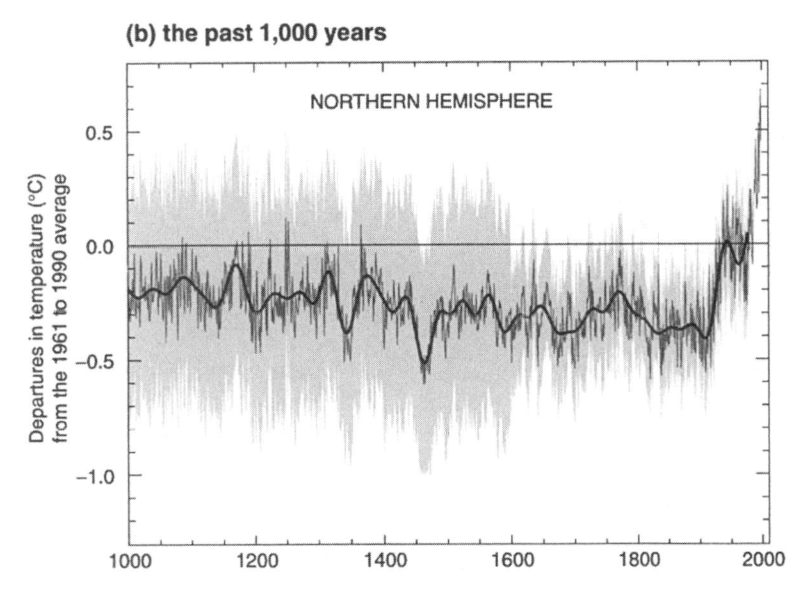

FIGURE I.6. Earth's Average Surface Temperature

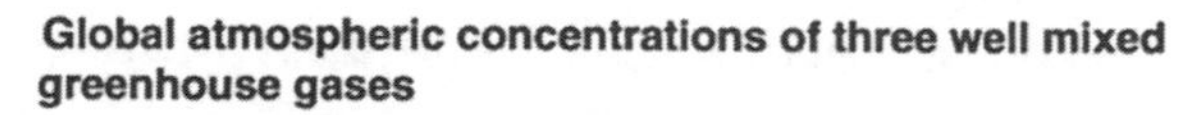

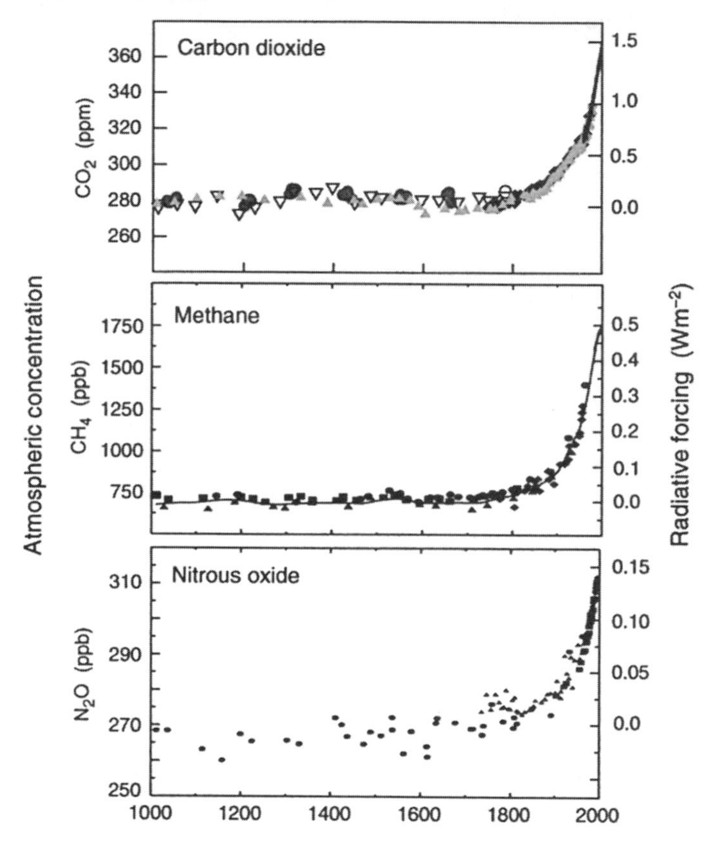

FIGURE I.7. Atmospheric Concentrations of Greenhouse Gases

Long-term cycles of atmospheric gases also must be investigated, to see if the current variations are part of some longer pattern. Using sediment records, a large cyclic variation in the carbon dioxide content is seen in the distant geologic past, but the data are incomplete and knowledge about the causes of the variations is lacking.

If the warming trend continues, many unfavorable consequences will result. Besides the obvious effects of sea level increases making some coastal regions uninhabitable as well as increasing salinity of freshwater lakes and streams, weather extremes would become more severe, with resulting loss of life and property damage. There would also be health difficulties: Tropical insects and diseases would migrate farther into temperate zones; diabetes, malaria, heat stroke, heat exhaustion, and breathing difficulties would all rise substantially.

As discussed in chapter 5, computerized climate models contain large uncertainties: modeling difficulties; solar variation; variable cloudiness; mathematical complexity; feedback due to interrelated, nonlinear climate variables; a grid size that is too large; and too little data. Just as in the case of weather, the final IPCC report was based on ensemble forecasting. It predicted severe impacts on human health, natural ecosystems, and agriculture and coastal communities, but acknowledged the large uncertainties involved.

An alternate and strongly held view is that the current global warming is simply part of some larger cycle that we do not fully understand, and any human actions have very little impact on this cycle.

Large-scale actions to reduce greenhouse gas emissions worldwide are still being studied, but the scientific uncertainties present decision makers with a cloudy picture—at least for now.

See the American Geophysical Union Web site: *www.agu.org/eos_elec/99148e.html.* For more up-to-the-minute news, do an Internet search on "greenhouse gases" or "global warming."

Future warming problems would be minimized if developed nations reduced fossil fuel use and extended renewable energy resources, such as hydro, wind, and solar. Nuclear energy is used in Europe, but its generation and use involve significant safety and waste issues. Further, third world nations need to reduce their birth rate. Ethical, economic, and political factors must all be considered before putting any plans into action.

11 Earth: The Inside Story

As planet Earth formed, gravity sorted the initial materials according to density, with the most dense materials sinking to the center and the least dense materials floating to become the crust. Figure I.8 presents a model of Earth's interior.

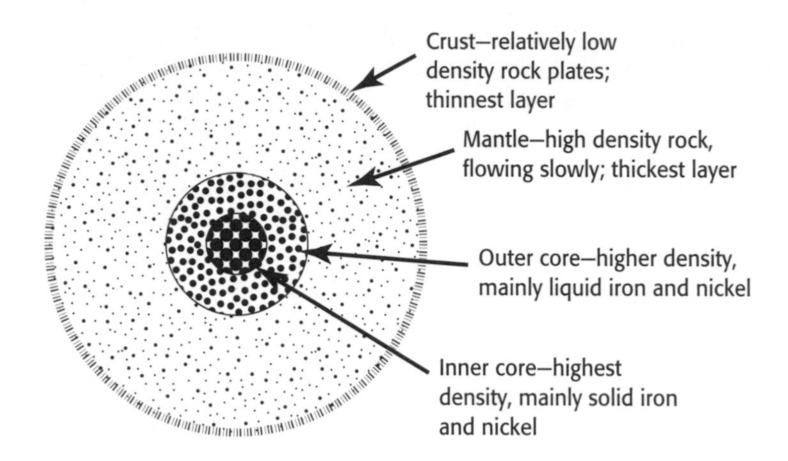

FIGURE I.8. Earth's Interior

The *crust* is the outermost layer. It is the least dense and is broken into many thin, rigid rock plates that move slowly because of underlying movements in the mantle.

The *mantle* is the next layer. It is the thickest of all layers, is relatively warm and fluid compared to the crust, and has hot spots that cause convection currents (think water undergoing a rolling boil, only much slower). Mantle currents drag the plates around, causing earthquakes, volcanoes, sea floor spreading, and continental drift.

The next layer is the hot, liquid *outer core.* It consists of dense iron and nickel that sloshes around because of Earth's rotation. Earth's magnetism may be caused by localized movement within this layer.

The innermost layer is called the *inner core.* While it also is made of hot iron and nickel, the materials are under so much pressure that the inner core is solid and the densest layer of all.

For more details about how this model was developed and the experimental evidence that supports it, see *The Five Biggest Ideas in Science* by Charles M. Wynn and Arthur W. Wiggins (New York: John Wiley & Sons, Inc., 1997).

The following Web sites contain up-to-date information and excellent graphics:

www.hartrao.ac.za/geodesy/tectonics.html

http://pubs.usgs.gov/publications/text/dynamic.html

www.seismo.unr.edu/ftp/pub/louie/class/100/plate-tectonics.html

http://scign.jpl.nasa.gov/learn/plate.htm

12 Chaos Theory

> Serious vanity! Mis-shapen chaos of well-seeming forms!
>
> —*William Shakespeare,* Romeo and Juliet

As discussed in chapter 5, chaos is not to be confused with randomness. Rather, chaos has come to mean extreme sensitivity of final results to small variations in initial conditions. As the old nursery rhyme goes,

For want of a nail, the shoe was lost;
For want of a shoe, the horse was lost;
For want of a horse, the rider was lost;
For want of a rider, the battle was lost;
For want of the battle, the kingdom was lost!

Prior to the 1960s, there was a purely mathematical development that turns out to be related to chaos theory. Gaston Maurice Julia (1893–1978), an Algerian mathematician, was wounded in World War I and had to wear a leather strap to protect his badly injured nose. Because he had to spend much of his time undergoing painful operations, he often conducted his mathematical research in hospitals. At age 25, he wrote "Memoire sur l'iteration des functions." A function is simply a mathematical rule for performing a calculation, such as $f(x) = x^2 + \text{constant}$. If x is 2 and the constant is 3, then the function's value is 7. Iteration is accomplished by using the value calculated for $f(x)$ as the new value for x. So, if $x = 7$, $f(x) = 52$. Julia studied more complicated expressions. He was especially interested in functions and values where the iteration could be repeated without having the result grow larger and larger without limit. Values of x for which repeated iterations produced finite results were termed "prisoners." If the results became infinitely large, they were called "escapers." The calculations were all carried out by hand and were quite laborious for even simple functions. Although Julia enjoyed a certain amount of fame in mathematical circles in the 1920s, his work was essentially forgotten until the 1970s.

Benoit Mandelbrot, born in Poland in 1924, was shown Julia's paper in 1945 by his uncle, a mathematics professor. Mandelbrot was not very interested in Julia's ideas at the time. However, 30 years later, after a highly vaned aca-

demic career, Mandelbrot wound up at IBM and applied computer power to Julia's iterative calculations. Further, Mandelbrot pioneered the graphical technique by which the computer displays the pattern of convergence and divergence of the function being iterated.

Figure I.9 shows a version of the resulting computerized plots.

The beautiful patterns generated by Mandelbrot and Julia techniques have graced countless books and Web sites for some time. For example:

Chaos: Making a New Science by James Gleick (New York: Viking Penguin, 1987)

Exploring Chaos—A Guide to the New Science of Disorder edited by Nina Hall (New York: W. W. Norton & Company, 1991)
http://hypertextbook.com/chaos/
www.wfu.edu/~petrejh4/chaosind.htm

In a related development, Steven Wolfram published *A New Kind of Science* in 2002 (see *www.Wolfram.com*). His work is based on personal research on cellular automata, which are an array of identically programmed automata, or "cells," that interact with one another according to fixed rules. Extremely complex images can be generated by very simple rules. Some images look very similar to examples in nature, but the connection between the mathematics of chaos and a useful description of the real world is a challenge that remains unresolved.

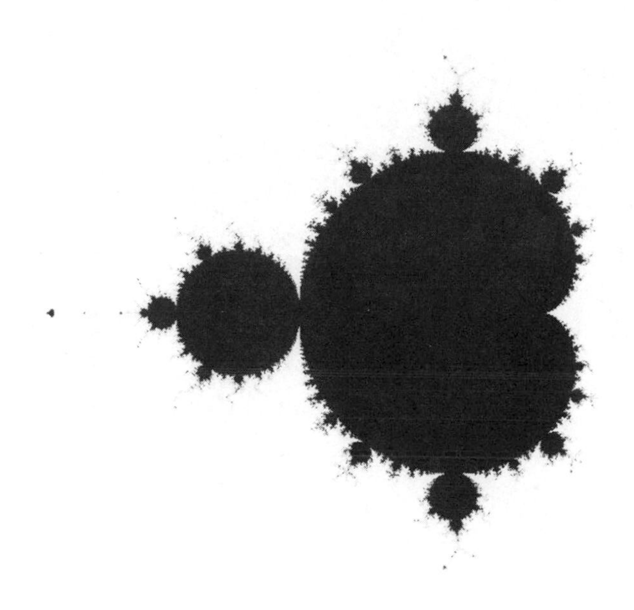

FIGURE I.9. Mandelbrot Design

13 Earthquake Prediction

Earthquake predictions are a dime a dozen. Internet search engines queried with "earthquake prediction" will turn up an average of more than 50,000 Web sites. Some earthquake predictions are made on the basis of psychic "readings." (See *Quantum Leaps in the Wrong Direction: Where Real Science Ends . . . and Pseudoscience Begins* by Charles M. Wynn, Arthur W. Wiggins, and Sidney Harris, Joseph Henry Press, Washington, D.C., 2001.) Other efforts focus on correlating earthquakes with ground electricity, animal behaviors, alignment of planets, or other effects. Although most predictions are wrong, every once in a while, one turns out to be right.

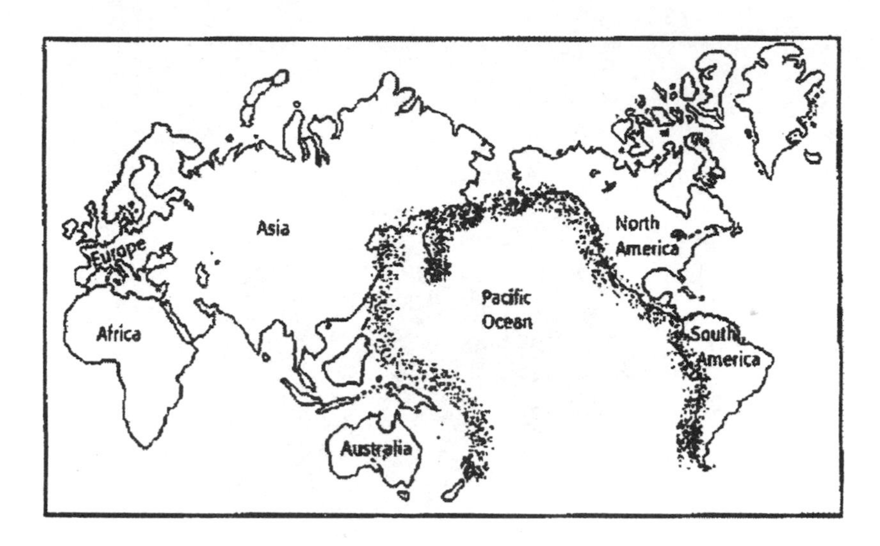

FIGURE I.10. Earthquake Zones

Suppose a friend offers you this wager. "I have $20 that says within the next month, a major earthquake will occur within the dotted area on the map above."

Don't take the bet. Your friend will probably win. The dotted area on the map in Figure I.10 corresponds to boundaries of the plates that make up Earth's crust. When mantle currents (see Idea Folder 11, Earth: The Inside Story) drag crustal plates around, the result is earthquakes. Although some earthquakes occur in areas other than at plate edges, that's where the vast majority happen. Earthquake magnitudes have the following statistical profile.

Magnitude on the Richter Scale (higher numbers are more severe)	Number of Earthquakes per Year
4–4.9	6,200
5–5.9	800
6–6.9	120
7–7.9	18

Note that the terms of the bet were quite loose. What constitutes a "major" earthquake? If your dividing line is a Richter Scale reading of 6 or above, more than 10 occur in an average month, most in the dotted area on the map. "Within a month" and somewhere within "the dotted area" are quite nonspecific terms. If you lived within that area, as millions of people do, should you evacuate? This prediction gives far too little information to be of any value.

In the 1970s, some geologists were optimistic about the possibility of specific and reliable earthquake prediction. There was even an extension of chaos theory called catastrophe theory, which looked as if it might be useful in predicting sudden events such as the buckling of beams, the cracking of cement slabs, or earthquakes. However, it turns out that constructing mathematical models of the dynamics of Earth's inner layers is almost as difficult as constructing atmospheric models. It's hard to write exact modeling equations, and even approximate equations are so nonlinear that they exhibit the sensitivity to initial conditions characteristic of chaotic systems. Also, obtaining information about the current state of the rock inside the crust and mantle is more difficult than measuring atmospheric variables, mostly because the interior of the crust and mantle is inaccessible.

In a 1997 *Science* article, prominent geologists Robert Geller, David Jackson, Yan Kagan, and Francesco Mulargia say, ". . . individual earthquakes are probably inherently unpredictable." The article's title says it all: "Earthquakes Cannot be Predicted." For details, see the Web site: *http://scec.ess.ucla.edu/~ykagan/perspective.html*

Here are some other good resources:

http://quake.wr.usgs.gov/research/parkfield/

www.nature.com/nature/debates/earthquake/equake_frameset.html

14 Compiling Star Catalogs

The following partial listing of star catalogs reflects the desire of humans to organize and search their surroundings for patterns. Even more ambitious plans are on the drawing boards for future observatories in space or on the Moon or Mars.

Stars are named according to the catalog in which they appear. Many bright stars are listed according to their names in Johann Bayer's catalog. Bayer assigned successive letters of the Greek alphabet to the brighter stars of each constellation. For example, Polaris is Alpha Ursae Minoris because it is the brightest (alpha) star in the constellation Ursa Minor. As another example, the first black hole's visible companion star was named HDE 226868 because it first appeared in the Henry Draper Extended catalog, so its location could be found by referring to 226868 there.

Date	Catalog Title and Star Designation	Compiled by	Number of Objects	Comments
350 BCE		Shih Shen	800	China
300 BCE		Timocharis		First true star catalog
130 BCE		Hipparchus	1,080	
120 CE	*Almagest*	Ptolemy (Claudius Ptolemaus)	1,022	See note 1
1540 CE	*De le Stelle Fisse*	Alesandro Piccolomini		48 Greek constellations
1602 CE		Tycho Brahe	about 1,000	See note 2
1603	*Uranometria* (listed as Greek letter + Latin constellation name)	Johann Bayer		Beautiful; positions from Brahe's data
1678		Edmund Halley		First southern sky catalog
1690	*Sternverzeichnis*	Johann Hevelius		Argued with Halley
1725	*Historia Coelestis Brittanica*	John Flamsteed	3,000	First Astronomer Royal; see note 3
1762		James Bradley	60,000	Third Astronomer Royal
1771	Nebulas called M	Charles Messier	100+	See chapter 6
1801		Johann Bode		Used earlier data
1860	*Bonner-Durchmusterung* (BD + CD + CPD)	Friedrich W. A. Argelander and others	1,160,000	Bonn Observatory
1864	*General Catalog* (GC)	Friedrich Wilhelm (William) Herschel, Caroline Herschel, John W. Herschel	2,500	See note 4
1888	*New General Catalog* (NGC and IC)	J. L. E. Dreyer	13,000	See note 5
1918–1924	*Henry Draper Catalog* (HD and HDE)	Edward C. Pickering and Annie Jump Cannon	400,000	See note 6

Date	Catalog Title and Star Designation	Compiled by	Number of Objects	Comments
1966	Smithsonian Astronomical Observatory (SAO)		260,000	Mt. Palomar plus
1989	*Hipparcos* (HIP) and *Tyco* (TYC)		2,500,000	Precision
1979-ongoing	*Guide Star* (GSC I and II)		1 billion	For pointing Hubble

Notes

1. Ptolemy's *Almagest* forms the basis for current astrological readings, even though Earth's axis has shifted (precessed) to the point that the zodiac constellations no longer line up in the appropriate months. Furthermore, a huge number of additional stars and even a few planets have been discovered since Ptolemy's time, all of which doesn't seem to matter to astrology.

2. Tycho Brahe, the last, greatest naked-eye observer, never published his star catalog. That task fell to Brahe's assistant, the worthy Johannes Kepler, who made his own more theoretical contributions to astronomy with his recognition of the orbital paths of planets as ellipses rather than circles.

3. John Flamsteed (1646–1719) founded the Royal Greenwich Observatory and became its first director and first Astronomer Royal. He was an exceedingly careful observer, who listed more stars and gave more accurate positions than any previous astronomer. Contemporaries Edmund Halley and Isaac Newton pressed Flamsteed through the Royal Society to release his observations early, even though they were incomplete. Despite his objections, 400 copies of partial results were published in 1712. Flamsteed managed to burn 300 copies. Nevertheless, Halley and Newton had done their work done quite well.

4. Sir William Herschel (1738–1822) was born Friedrich Wilhelm Herschel in Hanover, Germany. He moved to England in 1757, where he worked at musical jobs, including being an organist. In 1772, his sister Karoline Lucretia joined him in England. Shortly thereafter, his interest in astronomy grew, and his music pupils learned astronomy as well as music. With no room for a telescope at his house, he set one up on the street. It attracted visitors, one of whom was Dr. William Watson, member of the Royal Society, who presented some of Herschel's papers on the heights of Moon mountains to the society.

Within two years, Herschel found a bright object where prior charts showed no stars. This slow-moving object turned out to be a planet, named "Georgium sidus" by Herschel in honor of King George III, but later named Uranus. Herschel was elected Fellow of the Royal Society, and named court astronomer, with his sister as his assistant.

At age 50, Herschel married the widowed Mary Pitt. They had a son, John Frederick, who began his studies at Cambridge as a mathematician but later turned to astronomy and completed his father's catalog.

5. John Louis Emil Dreyer (1852–1926) was born in Copenhagen, Denmark. In 1872, he worked as an assistant to Lord Rosse at Birr Castle, between Dublin and Limerick in Ireland. Lord Rosse had built the world's largest telescope, a 72-inch-diameter monster nicknamed the Leviathan of Parsonstown. Although it had been shut down during the potato famine of 1845, when it was restarted, Dreyer observed many deep-sky objects and added 1,000 new entries to Herschel's *General Catalog*. Dreyer's major work came at Armagh Observatory, where he compiled the *New General Catalog* (NGC) at the request of the Royal Astronomical Society.

6. Henry Draper (1837–1882) was a medical doctor and amateur astronomer who, in 1872, made the first photograph of a stellar spectrum, that of the star Vega. After his untimely death at the height of his career, his widow endowed a fund that supported Harvard College Observatory's photographic stellar spectrum survey, headed originally by Edward C. Pickering. (See chapter 6).

After 1910, one of Pickering's "harem" (see chapter 6), Annie Jump Cannon, began classifying stars in terms of their spectra. She developed the scheme that divided the stars into spectral classes OBAFGKM (remembered by astronomy students as Oh, Be A Fine Girl (Guy), Kiss Me), and classified 50,000 stars a year, ultimately including nearly 400,000 stars and working for 40 years. In 1938, two years before her retirement, she obtained a regular Harvard appointment as William C. Bond astronomer.

For more information, see the Web site: *www.seds.org/~spider/spider/Misc/star_cats.html.*

15 Einstein's Works: Relativity Plus

Albert Einstein had five papers published in the German physics monthly journal *Annalen der Physik* in 1905:

In "A New Determination of Molecular Dimensions," Einstein showed how to obtain Avogadro's number (the celebrated 6.02×10^{23}, the number of molecules contained in one mole of a substance) and the sizes of ions in solution from measured values of osmotic pressure and the coefficient of diffusion. This work earned him his Ph.D. and is cited in papers written almost 100 years later.

"On the Motion—Required by the Molecular Kinetic Theory of Heat—of Small Particles Suspended in a Stationary Liquid" also dealt with molecules and explained how their jiggling motion seen under a microscope was caused by collisions with moving molecules in the liquid. The molecules themselves were too small to see, but the resulting motion of the larger body was visible to microscopists such as Robert Brown. This motion was called Brownian

motion. Einstein's paper reinforced the connection between kinetic theory and the reality of observations.

Einstein referred to his paper "On a Heuristic Viewpoint Concerning the Production and Transformation of Light" as being "revolutionary," and it was. Deeply dissatisfied with the description of matter as discontinuous as opposed to electromagnetic radiation's supposedly continuous nature,

Einstein suggested that light be treated as particles in some respects. He showed that this approach was consistent with Planck's analysis of the radiation of light from heated bodies. Applying this same logic to the photoelectric effect, where light shone on a metal causes electrons to be emitted, Einstein was able to explain several results that had baffled other theorists. This paper contributed to the new view of light by taking Planck's result more seriously than Planck himself, who regarded his treatment of light's discontinuous energy as more of a mathematical trick than an accurate representation of reality. By his own admission, Einstein had been thinking about this property of light for about five years before he wrote this paper.

"On the Electrodynamics of Moving Bodies" is Einstein's famous Special Theory of Relativity paper. It consists of a generalization of classical relativity, which states that the laws of physics are the same for any observer moving at constant speed. For example, if you toss a ball in the air inside a moving car, it goes up and comes down just as it would if you were standing on the ground. The second postulate of relativity is the revolutionary one. It reverses Newton's idea by making the speed of light constant for all observers moving at constant speed, while space and time are relative to the observer rather than being absolute, as in Newton's work. According to a letter he wrote to his uncle, Einstein had been thinking about this problem for at least seven years prior to 1905.

The last paper of 1905, "Does the Inertia of a Body Depend Upon Its Energy Content?" was a kind of mathematical footnote to the Special Theory of Relativity and contained the relationship between mass and energy. It was expressed in the form $m = L/V^2$ rather than the familiar $E = mc^2$.

For more details, see *Einstein's Miraculous Year: Five Papers That Changed the Face of Physics* edited by John Stachel, Princeton, NJ: Princeton University Press, 1998.

With such an enormous contribution in several areas of physics, you might think Einstein was extremely serious about his scientific pursuits. Here is the way he announced four of the papers to his friend Conrad Habicht in a letter sent in mid-1905:

> Such a solemn air of silence has descended between us that I almost feel as if I am committing a sacrilege when I break it now with some inconsequential babble. So, what are you up to, you frozen whale, you smoked, dried, canned piece of soul? Why have you still not sent me your dissertation? Don't you know that I am one of the 1.5 fellows who would read it with interest and pleasure, you wretched man? I promise you four papers in return.

> The first deals with radiation and the energy properties of light and is very revolutionary, as you will see if you send me your work first. The second paper is a determination of the true sizes of atoms. The third proves that bodies on the order of magnitude 1/1000 mm, suspended in liquids, must already perform an observable random motion that is produced by thermal motion. The fourth paper is only a rough draft at this point, and is on electrodynamics of moving bodies which employs a modification of the theory of space and time.

How did Albert Einstein manage to write five papers of such huge impact on physics in the short space of one year? You might be tempted to think Einstein was an enormously intelligent mathematical genius who was extremely successful in school, read widely, and worked in an academic setting that allowed him plenty of time for abstract thought. Not exactly.

In 1905, Albert Einstein was 26 years old, worked full time at the Swiss Patent Office in Bern, was married to Mileva Maric (1875–1948), his college sweetheart, and was the father of a one-year-old son, Hans Albert.

Here are a few quotes from Albert Einstein that illustrate his thinking:

> I have no particular talent. I am merely inquisitive.

> It's not that I'm so smart, it's just that I stay with problems longer.

> These thoughts did not come in any verbal formulation. I rarely think in words at all. A thought comes, and I may try to express it in words afterward.

Once Einstein sent this reply, along with a page full of diagrams, to a 15-year-old girl who had written for help on a homework assignment:

> Do not worry about your difficulties in mathematics; I can assure you that mine are much greater.

> I sometimes ask myself how it came about that I was the one to develop the theory of relativity. The reason, I think, is that a normal adult never stops to think about problems of space and time. These are things which he has thought about as a child. But my intellectual development was retarded, as a result of which I began to wonder about space and time only when I had already grown up.

Many biographers chronicle Einstein's early school career, with his history of independence, unwillingness to follow authority, and numerous failures. Some have theorized that he had a learning disability, possibly dyslexia.

Another quote may bear on this question: "Reading, after a certain age, diverts the mind too much from its creative pursuits. Any man who reads too much and uses his own brain too little falls into lazy habits of thinking."

Certainly Einstein's intelligence was far above average, but perhaps his ability to concentrate was more important. Some people would call it stubbornness, but focusing his considerable abilities on one topic served him well. Yet his intense interest in science probably didn't make him an ideal husband and father. After the considerable fame brought by his scientific work, Einstein accepted several academic posts and traveled extensively. This all took its toll, and he and Mileva divorced in 1919. As part of the divorce settlement, Einstein agreed to turn over the cash part of any future Nobel Prize for support. The Nobel Prize was awarded to Einstein in 1921 (for the photoelectric effect explanation), and his ex-wife and children got the money.

Albert Einstein married his cousin's widow, Elsa, in 1919 and continued his scientific work and travel, playing his violin (he called it a fiddle) all over the world. Although few understood his theories, the language of music appealed to all. In 1919, the first experimental support for his general relativity theory was produced, bringing him still more fame. The rise of the Nazis in Germany created more and more trouble for the pacifistic and Jewish Einstein. He eventually fled to the United States. At Princeton's Institute for Advanced Studies, he pursued a unified field theory fruitlessly. Einstein served as physics' elder statesman until his death in 1955.

> The world needs heroes and it's better they be harmless men like me than villains like Hitler.
>
> *—Albert Einstein*

16 The Big Bang

The Big Bang theory of the universe says that all the matter and energy in the universe started out as a single point about 14 billion years ago and then began expanding. During its early stages, it expanded extremely rapidly, in a

"IT WAS A LOT EASIER TO KEEP AN EYE ON THINGS BEFORE THE BIG BANG. EVERYTHING WAS ALL IN ONE PLACE THEN."

period called inflation, then it continued its expansion in a decelerating fashion because of the influence of gravity. Now it's accelerating again because of dark energy.

For a more detailed treatment, including experimental support, see *The Five Biggest Ideas in Science* by Charles M. Wynn and Arthur W. Wiggins (New York: John Wiley & Sons, 1997).

Resources for Digging Deeper

General Resources

Books

Anton, Ted. *Bold Science; Seven Scientists Who Are Changing Our World.* New York: W. H. Freeman and Co., 2000.

Kaku, Michio. *Hyperspace.* London: Oxford University Press, 1994.

Kaku, Michio. *Visions.* New York: Anchor Books, 1997.

Kuhn, Robert L. *Closer to Truth: Challenging Current Belief.* New York: McGraw-Hill, 2000.

Periodicals

Discover

Science

Science Week

Scientific American (or *www.sciam.com*)

Web Site

www.mkaku.org

Chapter 1 Science in Perspective

Books

Brockman, John (Ed.). *The Next Fifty Years—Science in the First Half of the Twenty-First Century.* New York: Vantage Books, 2002.

Malone, John. *Unsolved Mysteries of Science: A Mind-Expanding Journey through a Universe of Big Bangs, Particle Waves, and Other Perplexing Concepts.* New York: John Wiley & Sons, Inc., 2001.

Chapter 2 Physics: Why Do Some Particles Have Mass While Others Have None?

Books

Brennan, Richard P. *Heisenberg Probably Slept Here: The Lives, Times, and Ideas of the Great Physicists of the 20th Century.* New York: John Wiley & Sons, Inc., 1996.

Kane, Gordon. *Supersymmetry: Squarks, Photinos, and the Unveiling of the Ultimate Laws of Nature.* Cambridge, Mass.: Helix Books, 2000.

Peat, F. David. *Superstrings and the Search for the Theory of Everything.* New York: Contemporary Books, 1989.

Periodicals

Arkani-Hamed, Nina, Savas Dimopolous, and Georgi Dvali. "The Universe's Unseen Dimensions." *Scientific American,* August 2000.

"A Matter of Time," *Scientific American,* September 2002 Special Issue.

Overbye, Dennis. "Remembering David Schramm, the Gentle Giant of Cosmology." *New York Times,* February 10, 1998.

Weinberg, Steven. "A Unified Physics by 2050?" *Scientific American,* December 1999.

Web Sites

CERN (The European High Energy Physics site): *http://welcome.cern.ch/welcome/gateway.html*

Contemporary Physics Education Project: *www.cpepweb.org/*

Fermi National Accelerator Laboratory site: *www.fnal.gov/*

Higgs field: *www.hep.yorku.ca/what_is_higgs.html*

Higgs search: *http://magazine.uchicago.edu/0104/features/higgs.html*

High Energy Physics at Fermilab: *www.hep.net/*

"Hunting for Higher Dimensions," *Science News Online,* February 19, 2000: *www.sciencenews.org*

"A Layman's Guide to M-Theory" by M. J. Duff, *http://arxiv.org/abs/hep-th/9805177*

Particle Adventure site: *http://particleadventure.org/particleadventure/index.html*

Particle Physics and Astronomy Research Council site: *www.pparc.ac.uk/*

Quantum Field Theory: *http://theory.caltech.edu/people/jhs/strings/str114.html*

Stanford Linear Accelerator Center site: *www.slac.stanford.edu/*

Chapter 3 Chemistry: By What Series of Chemical Reactions Did Atoms Form the First Living Things?

Books

Adams, Fred. *Origins of Existence: How Life Emerged in the Universe.* New York: The Free Press, 2002.

deDuvé, Christian. *Life Evolving: Molecules, Mind, and Meaning.* Oxford: Oxford University Press, 2002.

Ridley, Matt. *Genome.* New York: HarperCollins, 2000.

Shapiro, Robert. *Planetary Dreams: The Quest to Discover Life Beyond Earth.* New York: John Wiley & Sons, Inc., 2001.

Periodicals

Ridley, Matt. "The Year of the Genome." *Discover,* 22, No. 1, January 2001.

Wade, Nicholas. "Inside the Cell, Experts See Life's Origin." *New York Times,* April 6, 1999.

Web Sites

Archaea: *www.ucmp.berkeley.edu/archaea/archaea.html*

Beginnings of life on Earth: *www.sigmaxi.org/amsci/articles/95articles/cdeduve.html*

Life in the right universe: *www.discover.com/nov_00/featlife.html*

Origin of life: *http://origins.jpl.nasa.gov/*
www.resa.net/nasa/origins_life.htm
http://taggart.glg.msu.edu/isb200/oolife.htm

The Origin of Life on Earth, by Leslie Orgel:
www.geocities.com/CapeCanaveral/Lab/2948/orgel.html

Origin of Life prize: *www.us.net/life/*

Origins and early evolution of Life: www.chemistry.ucsc.edu/Projects/origin/home.html

Wickramasinghe and Hoyle's view of life's origins: www.actionbioscience.org/newfrontiers/wickramasinghe/wick_hoyle.html

Chapter 4 Biology: What Is the Complete Structure and Function of the Proteome?

Book

Raven, Peter H., and George B. Johnson. *Biology, 6th Edition.* New York: McGraw-Hill, 2002.

Web Sites

Applied molecular genetics: *www.biochem.arizona.edu/classes/bioc471/pages/Lecture3.html*

Biochips: *http://157.98.13.103/docs/1995/103-3/innovations.html*
http://arrayit.com/Company/Media/PrintMedia/printmedia.html
www.goertzel.org/benzine/FodorProfile.htm

Gel electrophoresis: *www.iacr.bbsrc.ac.uk/notebook/courses/guide/dnast.htm*

Genetic code: *http://newton.dep.anl.gov/askasci/mole00.htm*

"Junk" DNA, or is it?: *www.iacr.bbsrc.ac.uk/notebook/courses/guide/dnast.htm*

Molecular biology notebook: *www.iacr.bbsrc.ac.uk/notebook/courses/guide/dnast.htm*

Molecular genetics: *http://newton.dep.anl.gov/askasci/mole00.htm*

Chapter 5 Geology: Is Accurate Long-range Weather Forecasting Possible?

Periodical

"Scientific American Presents Weather," *Scientific American,* 11, No. 1, Spring 2000.

Web Sites

Ice on the Moon: *http://nssdc.gsfc.nasa.gov/planetary/ice/ice_moon.html*

Origin of water on Earth: *http://scienceweek.com/swfr065.htm*

Running climate prediction simulations on your computer: *www.climateprediction.com/*

Venera missions to Venus: *http://nssdc.gsfc.nasa.gov/planetary/venera.html*

Chapter 6 Astronomy: Why Is the Universe Expanding Faster and Faster?

Books

Bergstrom, L., and A. Goobar. *Cosmology and Particle Astrophysics.* New York: John Wiley & Sons, Inc., 1999.

Boss, Alan. *Looking for Earths: The Race to Find New Solar Systems.* New York: John Wiley & Sons, Inc., 2000.

Fox, Karen C. *The Big Bang Theory: What It Is, Where It Came From, and Why It Works.* New York: John Wiley & Sons, Inc., 2002.

Livio, Mario. *The Accelerating Universe: Infinite Expansion, the Cosmological Constant, and the Beauty of the Cosmos.* New York: John Wiley & Sons, Inc., 2000.

Periodicals

Cline, David B. "The Search for Dark Matter." *Scientific American* 288, 3, March 2003.

Overbye, Dennis. "A Scientist's Prey: Dark Energy in the Cosmic Abyss." *New York Times,* February 18, 2003.

Wright, Karen. "Very Dark Energy," *Discover,* 22, No. 3, March 2001.

Web Sites

Acceleration of the universe: *www.discover.com/science_news/astronomy/quick.html*

Astronomy links: *www.winternet.com/~gmcdavid/html_dir/astronomy.html*

Friedrich Bessel biography: *www.groups.dcs.stand.ac.uk/~history/Mathematicians/Bessel.html*

Big Bang plus: *http://hoku.as.utexas.edu/~gebhardt/a309s02/lect5dm.html*

Big Bang theory: *www.damtp.cam.ac.uk/user/gr/public/bb_home.html*

Cosmological constant and dark matter: *http://umwnt1.physics.lsa.umich.edu/PIC99/_Talks/turner/turner.htm*

Dark energy in the accelerating universe: *http://snap.lbl.gov/brochure/index.html*

Dark energy resource book: *http://supernova.lbl.gov/~evlinder/sci.html#sec1*

Dark matter and dark energy: *http://hitoshi.berkeley.edu/290E/*

High Z Supernova Project: *www.sc.doe.gov/feature_articles_2001/April/lucky_supernova/lucky_supernova.htm*

Microwave Anisotropy Project: *http://map.gsfc.nasa.gov/m_uni/uni_101fate.html*

M-Theory: *www.damtp.cam.ac.uk/user/gr/public/qg_ss.html*

Next Generation Space Telescope: *http://ngst.gsfc.nasa.gov/*

Supernova acceleration probe presentation: *http://atlas.physics.lsa.umich.edu/docushare/dscgi/ds.py/GetRepr/File-985/html*

Theoretical cosmology links: *http://cfa-www.harvard.edu/~jcohn/tcosmo.html*

Problem Folders

Books

Kaku, Michio. *Hyperspace.* New York: Oxford University Press, 1994.

Kaku, Michio. *Visions.* New York: Anchor Books, 1997.

Malone, John. *Unsolved Mysteries of Science.* New York: John Wiley & Sons, Inc., 2001.

Penrose, Roger. *The Emperor's New Mind: Concerning Computers, Minds, and the Laws of Physics.* New York: Viking Penguin, 1990.

Raup, David. *Extinction: Bad Genes or Bad Luck?* W. W. Norton & Company, New York, 1992.

Rees, Martin. *Our Cosmic Habitat.* Princeton, NJ: Princeton University Press, 2001.

Steel, Duncan. *Rogue Asteroids and Doomsday Comets: The Search for the Million Megaton Menace That Threatens Life on Earth.* New York: John Wiley & Sons, Inc., 1997.

Periodicals

Crick, Francis, and C. Koch. "The Problem of Consciousness," *Scientific American,* September 1992.

Gibbs, W. Wayt. "Ripples in Spacetime," *Scientific American,* April 2002.

Overbye, Dennis. "A New View of Our Universe: Only One of Many," *New York Times,* October 29, 2002.

Wade, Nicholas. "Before the Big Bang, There Was . . . What?" *New York Times,* May 23, 2001.

Web Sites

www.jupiterscientific.org/sciinfo/gusp.html

www.mkaku.org

http://neat.jpl.nasa.gov

http://neo.jpl.nasa.gov

http://spacewatch.lpl.arizona.edu

Photo Credits

Page 81: Courtesy James Priess, Fred Hutchinson Cancer Research Center; page 94: AP/Worldwide Photos; page 111: Courtesy Jim Bell (Cornell U.), Steve Lee (U. Colorado), Mike Wolff (SSI), STScI/NASA; page 112: Courtesy Nick Strobel/www.astronomynotes.com; pages 123 top, 205: Courtesy Gary William Flake, *The Computational Beauty of Nature;* page 132: Courtesy STScI/NASA; page 134: Gustav Giessler/Courtesy of the Albert Einstein Archives, The Jewish National & University Library, Hebrew University of Jerusalem, Israel; page 138: The University of Chicago, Yerkes Observatory; page 139: Courtesy Margaret Harwood, AIP Emilio Segré Visual Archives, Shapely Collection; page 141: Courtesy AIP Emilio Segré Visual Archives, Shapely Collection; page 142: Courtesy Observatory Collection, Photographs Box 17, Benteley Historical Library, the University of Michigan; page 143: Courtesy AIP Emilio Segré Visual Archives; page 147: Courtesy the Carnegie Observatories, Carnegie Institution of Washington; page 149: Courtesy of the Archives, California Institute of Technology; page 155: Courtesy the Boomerang Collaboration; page 176: Photography by Watson Davis, Science Service, Berkeley, Courtesy AIP Emilio Segré Visual Archives, Fermi Film; page 187: Courtesy Barbara M. Wiggins; pages 199, 200: Courtesy of the Intergovernmental Panel on Climate Change.

Index

A

ABI 373A Sequencer, 89, 90, 91
ABI 800 Catalyst workstation, 89, 91
ABI PRISM 3700, 92
actin, 191
action-at-a-distance problem, 24, 25
agriculture, and DNA applications, 98
albedo, 104–105
Alpha Centauri, 131
alpha particles, 7, 15, 22
Altman, Sidney, 55
Alvarez, Luis and Walter, 164
amino acids, 46–47, 54, 59, 60, 75, 185–186, 188–190
  in animals, 186
  and chirality, 65, 66
  and proteins, 191–192
Anaxagoras, 43
Anderson, Carl D., 17, 172, 174
Andromeda (M31) galaxy, 132, 141, 142, 144–145
annihilation, 172, 173
anti-atoms, 172
anti-electrons, *see* positrons
anti-leptons, 25
anti-matter, 172–174
anti-mesons, 174
anti-muons, 17
anti-neutrons, 172
anti-particles, 17, 22, 172
anti-protons, 172
anti-quarks, 22, 25
Applied Biosystems, Inc., 89
archaea, 49–50
ARGO, 124
Aristotle, 102
Arrhenius, Svante, 44
Astbury, W. T., 191
asteroid impacts, 169
astronomy, 127
atoms, 27
  anti-, 172
  in early universe, 39
  models of, 7, 16
  structure of, 7, 14, 15
  workings of, 27
axions, 150, 156–157

B

Baade, Walter, 147–148
background microwave radiation, 154, 155, 157
bacteriophages, 85–86
baryonic dark matter, 148–150
baryons, 22
Bayer, Johann, 207, 208
Beijing Genomics Institute, 96
Bessel, Friedrich, 130
Besso, Michele Angelo, 134
beta-galactosidase, 73–74
Big Bang, 28–29, 38–39, 146, 154, 170, 214–215

Big Crunch, 152
biochips, 97
bioinformatics, 99
biology, 71, 83, 84
Bjorken, James, 23
black holes, 10, 29, 150, 160–161
blueshift, 144
B mesons, 174
Bohr, Niels, 35
bombardment, of planets, 114–117
BOOMERANG Project, 155
Bose, Satyendranath, 28, 178
Bose-Einstein condensate, 178
Bose-Einstein statistics, 178
bosons, 25, 27, 28, 30, 172, 177, 178
Brahe, Tycho, 208, 209
brane theory, *see* multidimensional membranes
Brenner, Sydney, 81–82, 90
brewer's yeast, 80–81, 91, 96
Brookhaven National Laboratory, 6
brown dwarfs, 149–150
Burlingame, Alma L., 100
butterfly effect, 122–123

C

Caenorhabditis elegans, 81–82
Cairns-Smith, A. G., 56, 64
calculus, 35
cancer, 163, 195
Cannon, Annie Jump, 208, 210
carbonate rocks, 119
carbonate-silicate cycle, 119
carbon-based life, 39, 179–180
carbon compounds, 39
carbon dioxide, 197, 198
Castle, William, 82
catabolite activator protein, 75–78
catalysts, 52–53, 62
catastrophe theory, 126, 207
Cech, Thomas, 55
Celera, 92, 99
cell communication, 163
cells, 15, 49–50
  aging of, 163
  functions of, 50–56
Center for Human Genome Research, 88, 91
Center of the Advancement of Genomics, 196
Cepheid variables, 139, 142, 145
cereal grains genes, 96
CERN (European Organization for Nuclear Research), 31, 36, 176–177
Chamberlain, Owen, 172
chaos theory, 103, 123, 203–204
chaperonins, 193
chemical reactions, and atmospheric gases, 115–117
chemistry, 37, 39
chirality, 65, 66
circular accelerator, *see* cyclotron
Clarke, Arthur C., 184
Clay World hypothesis, 56, 64
climate, 120, 165–166, 198
Clinton, President Bill, 93, 94
cloning, 98
closed universe, 151, 155
coacervates, 45
codons, 75, 188–190
cold dark matter, 150
collagen, 191
Collins, Francis, 91, 92, 93, 94
Coma Berenices cluster, 148
Comet Wild 2, 67
comets, 58, 114, 132
Committee for the Physics of the Universe, 158
complementary DNA (cDNA), 90
complex numbers, 34–35
condensation, 55, 115
  on Mars, 117
  on Venus, 116
conformational change, 75
consciousness, human, 165
Corey, Robert C., 191
corn genome, 96
correspondence principle, 35
cosmic dust, 66, 67
cosmic rays, 17, 173–175
cosmological constant, 136, 146, 156
cosmology, 127

Cowan, Clyde, 19
cratering, of planets, 115–117
Cray, Seymour, 125
Crick, Francis, 47, 83, 189, 191
critical density, 152
critical universe, 151
Curien, Hubert, 90
Curtis, Heber D., 140, 142
cyclic adenosine monophosphate (cAMP), 75–77
cyclotrons, 19, 21, 175
Cygnus X-1, 9–10

D

D'Herelle, Felix, 85
dark energy, 127, 129, 154–158
dark matter, 129, 148–150, 152, 155, 156
DASI Project, 155
deductive reasoning, 5
deDuvé, Christian, 181
deforestation, 198
Deimos, 110
denaturation, 192
deoxyribonucleic acid, *see* DNA
deoxythymidine triphosphate (dTTP), 87
descriptive biology, 83, 84
determinism, 120
deuterium, 23
dideoxy bases, 87–88
differentiation (gravitational separation), 113
dimer, 55
dim matter, 149
dinosaurs, 164–165
Dirac, P. A. M., 16, 17, 28, 172, 178
disaccharides, 73
diseases, and genetic technologies, 194
dissociation, 115
DNA (deoxyribonucleic acid), 48–55, 74, 83
  altering of in plants, 98
  and chirality, 65
  and codons, 188
  information storage in, 63
  models of, 187–188
  and protein enzymes, 86
  and RNA, 75–77
  sequence of bases of, 84–88
  see also complementary DNA
DNA sequencing, 86–95, 196
Doppler effect, 144–145
Drake, Frank, 180, 184
Drake equation, 180–182
Draper, Henry, 207, 208, 210
Dreyer, John L. E., 208, 210
Drosophila melanogaster, see fruit flies
Dulbecco, Renato, 88

E

Earth, 118
  age of, 8
  and asteroid impacts, 169
  atmosphere of, 102–104, 111–116, 118–120, 196–201
  carbon dioxide on, 116, 119
  climate changes on, 165–166
  crust of, 202
  escape velocity of, 115
  formation of, 58
  and greenhouse gases, 196–201
  interior of, 166, 201–202
  oxygen on, 119
  retention of water on, 118–119
  surface temperature of, 198, 199
earthquake prediction, 166, 205–207
earthquake zones, 206
Eidgenössische Technische Hochschule (ETH), 133, 135
Einstein, Albert, 5, 16, 28, 128, 133–137, 146, 147, 156, 178, 210–214
  relativity theories, 8, 16, 33, 134, 135, 212
electric field, 25
electricity, and magnetism, 2, 23
electromagnetic interaction, 28
electromagnetism, 2, 23
electron neutrino, 25, 26, 27
electrons, 7, 14, 16, 25–27, 172
electron scattering, 23
electron spin, 177–178

electrophoresis, 84–85, 88
electroweak interaction, 23, 25
electroweak symmetry breaking, 30
element 118, 6
elements, synthesis of, 6
El Niño, 124
energy, and mass, 128
ensemble forecasting, 125
enzymes, 52, 55
see also restriction enzymes
Equivalence Principle, 134
escape velocity, 115
Escherichia coli (*E. coli*), 73, 75, 76, 78, 79
ethics, 3, 98–99
eukaryotes, 49, 50, 79–81
European Organization for Nuclear Research, *see* CERN
European Space Agency, 107, 111, 157
evaporation, and planet atmospheres, 114, 116, 117
experiment, 4–8, 11
experimental evidence, 8
Expressed Sequence Tags, 90, 91
extinctions, mass, 164
extraterrestrial life, 167, 179–185
extremophiles, 49

F
falsifiability, 5
Faraday, Michael, 2
Fermi, Enrico, 19, 22, 28, 178, 183
Fermi-Dirac statistics, 178
Fermi National Accelerator Laboratory (Fermilab), 31, 36, 176
fermions, 26, 28, 177
Fermi's Paradox, 183
Feynman, Richard P., 23
fields, 24–25
see also Higgs field
finite difference analysis, 122
FitzGerald, George, 7
Flamsteed, John, 208, 209
flat universe, 152, 155
Fodor, Stephen, 97
forces (in nature), 17–18
and fields, 24
see also electromagnetism; gravity; strong nuclear force; weak nuclear force
Ford, W. K., 148
Fox, Sidney, 56, 64
Franklin, Rosalind, 83
Frayn, Michael, 20
Fritszch, Harald, 24
fruit flies, 82, 93
functions, mathematical, 203
fusion, nuclear, 56

G
galactose, 73, 74
galaxies, 131–133, 140–141
distances to, 145
formation theories, 167
Hubble classification of, 143–144
speeds of, 144–145
gamma-ray bursts, 167–168
Geiger, Hans, 7
Gell-Mann, Murray, 22, 23–24
General Theory of Relativity, 33, 134–135
genes, 62–63
in human genome, 95
patenting of, 90–91
genetic technologies, 193–196
genomics, 97–99
geology, 101
Giessler, Gustav, 134
Gilbert, Walter, 55, 88
Gladstone, William, 2
Glashow, Sheldon, 23
Gleick, James, 125
Global Climate Observation System, 124
global warming, 165–166, 198–201
globular clusters, 140
globular proteins, 190
glucose, 73–74, 76–77
gluons, 24, 25, 27, 28, 172, 178
Gold, Thomas, 65
Goudsmit, Samuel, 177
grand unification theory (GUT), 33–34
gravitation, 135, 136
gravitational field, 24

gravitational theory, 33, 150–151
gravitational waves, 162
graviton, 172
gravity, 33
  and mass, 30
Green Bank equation, *see* Drake equation
greenhouse effect, 106, 116, 117
greenhouse gases, 119, 196–201
Grossmann, Gerald, 134
Grossmann, Marcel, 133–135

H

Habicht, Conrad, 134
hadrons, 22
Haemophilus influenzae, 79, 86, 92
Haldane, J. B. S., 45–49, 67
Hale, George Ellery, 137–138
Hale telescope, 138
Halley, Edmund, 209
Harvey, William, 12
Hawkins, Benjamin W., 164
HDE 226868 (star), 9–10, 41
Healy, Bernadine, 91
Heisenberg Uncertainty Principle, 16, 23
helium, in universe, 29, 38
Helmholtz, Hermann von, 44
Helmont, Jan Baptiste, 41
hemoglobin, 190, 192
Herschel, Caroline, 208, 209
Herschel, Friedrich W., 208, 209
Herschel, John Frederick, 209
Hertzsprung, Ejnar, 140
hierarchy problem, 33
Higgs, Peter, 31
Higgs boson, 27, 177
Higgs field, 14, 31, 33
Higgs field problem, *see* mass: origin of problem
Higgs particle, 23, 28, 31–33, 36
High-z Supernova Team, 153
Hood, Leroy, 89, 92
Hooker, John D., 137
Hooker telescope, 137, 141–143
Horvitz, Robert, 82
hot dark matter, 150
Hoyle, Fred, 64, 66, 146
Hoyle-Wickramasinghe theory, 66, 67
Hubble, Edwin, 137, 141–147
Hubble's law, 145–146
human genetic control, 98
human genome, 71, 84, 95, 97, 196
  sequencing of, 88–95
Human Genome Project, 88–95
Human Genome Sciences (HGS), 91, 92
human sciences, 3, 4, 5
Humason, Milton, 145, 147
Hungerford, Margaret Wolfe, 3
Hunkapiller, Michael, 89, 92
hydrologic cycle, 118
hydrothermal vents, 59
hyperthermophiles, 49
hypotheses
  multiple, 9–10
  and Ockham's Razor, 9–10
  and scientific method, 4–5, 7, 8, 10–11

I

ice ages, 165
inductive reasoning, 4
initial conditions, 120–121
Institute for Biological Energy Alternatives, 196
Institute for Genomic Research, The (TIGR), 91, 92, 196
Intergovernmental Panel on Climate Change (IPCC), 198, 201
International Rice Genome Sequencing Project (IRGSP), 96
island universes, 132, 142
iteration, 203

J

Jacob, François, 73
James Webb Space Telescope, 157
Julia, Gaston Maurice, 203, 204
Jupiter, 104

K

K-12 strain (*E. coli*), 73, 79
Kant, Immanuel, 131–132
kaons, 21
Kendrew, John C., 191

Kepler, Johannes, 209
ketone, 65
K meson, 174

L

La Niña, 124
lac operon, 73, 76, 78, 79, 80
lac repressor, 75–77
lactose, 73–74, 75–77
Lamb, Willis, 28
Lamb shift, 28
lambda particle, 21
Lane, Neal, 93
Large Electron Positron Collider, 31, 177
Large Hadron Collider, 31, 177
Large Magellanic Cloud, 139
Lawrence, Ernest O., 19, 175, 176
Leavitt, Henrietta Swan, 137, 138–140
Lederman, Leon, 22
leptons, 22, 25, 26, 27
life
  dark matter/dark energy-based, 185
  origin hypotheses, 40–48
  origins of, 37–69
  see also extraterrestrial life
life-forms
  carbon-based, 39, 179–180
  silicon-based, 180
light, 16, 160
Linderstrøm-Lang, K., 191
linear accelerators, 175
Livingston, M. Stanley, 175
Lorenz, Edward, 122
Lunar Prospector, 115
Lwoff, André, 73

M

M31 (Andromeda) galaxy, 132, 141, 142, 144–145
M100 (Whirlpool) galaxy, 132
MACHOs (Massive Compact Halo Objects), 149–150
magnetic field, 25
magnetic moment of the muon, 6
magnetism, and electricity, 2, 23
  see also electromagnetism
Mandelbrot, Benoit, 203–204
Mandelbrot Design, 204, 205
Manhattan Project, 20
Mars, 108–111, 118
  atmosphere of, 104, 109–115, 117, 118
  cratering on, 117
  escape velocity of, 115
  ice on, 110, 118
  water on, 117–118
Marsden, Ernest, 7
Mars Express, 111
mass, 14
  and energy, 128
  and gravity, 30
  origin of, 13–36
  origin of problem, 31–33
mass/energy density, 155
mass-energy equivalence, 27
mass extinctions, 164
mathematics, and hypotheses, 5
matter, 14, 129, 155, 156
  see also anti-matter
matter/energy density, 152
MAXIMA Project, 155
Maxwell, James Clerk, 8, 23
Mercury, 104, 111, 112, 114
mesons, 22, 174
messenger RNA (mRNA), 188
Messier, Charles, 132–133
meteorites, and organic molecules, 66–67
methane, 44–45, 197–198
methylase, 86
mice (*Mus musculus*), 82
microspheres, 64
Milky Way galaxy, 131, 140
Miller, Stanley L., 46–47, 48
Millikan, Robert A., 147
model organisms, 81–83
molecular biology, 83, 84
molecules, 15, 39
  handedness of, 65
  shapes of, 163
Monod, Jacques, 73, 79
monomers, 55
Moon, 104, 114, 115
Morrison, Philip, 184

Mount Wilson Observatory, 137
M-theory, 34, 174
multidimensional membranes ("branes"), 34, 169, 174
multiple universes, 157, 169
muon neutrinos, 25, 26, 27
muons, 17, 25, 26, 27
  magnetic moment of, 6
Mycoplasma genitalium, 79, 92
myosin, 191

N
National Institutes for Health, 88, 89, 91
natural sciences, 3, 4, 5
Near Earth Asteroid Tracking (NEAT), 169
Near-Earth Objects (NEOs), 169
nebulae, 131–133
Neptune, 104
Neumann, John von, 122
neutrinos, 19, 25–27, 129, 150
neutrons, 14, 15, 16, 27, 172
  and quarks, 22–23
neutron stars, 10, 150
Newton, Isaac, 5, 120, 209, 212
Nirenberg, Marshall, 189
nitrous oxide, 198
nuclear fusion, 56
nuclear winter, 110
nucleosides, 60
nucleotide base pairs, 95–96
nucleotide bases, 188–189
nucleotides, 50, 52, 54, 60, 62–63
nucleus (of atom), 7, 15, 27
numbers, complex, 34–35

O
objectivity, and science, 3, 8
observation, 4, 7, 10
Ockham's Razor, 9, 10
Oelert, Walter, 172
Omega ratio, 152
Oparin, Aleksandr I., 44–46
Oparin-Haldane theory, 46–48, 67
open universe, 151, 155
operons, 73
  see also lac operon
organic chemistry, 39
Orgel, Leslie, 55
Origin-of-Life Prize, 69
outgassing, 114, 116, 117
Owen, Richard, 164
oxygen, on Earth, 119
ozone, 119
ozone layer, 60

P
panspermia, 43–44, 65–66, 67
parallax, 129–130
parameters, 30
parsecs, 130–131
particle accelerators, 19, 20–21, 23, 174–177
particle decay, 27
particles
  discovery of new, 17, 19, 20–22
  see also anti-particles
Pasteur, Louis, 42–43, 69
Patrinos, Ari, 93
Pauli, Wolfgang, 178
Pauli exclusion principle, 178
Pauling, Linus, 191
peptide nucleic acid, 61
Period-Luminosity relation, 139–140
permease, 73
perturbation, 30
phage P22, 86
phages, *see* bacteriophages
Phobos, 110
phosphate, 59, 60
photinos, 150
photodissociation, 60
photoelectric effect, 16
photons, 23, 25, 27, 28, 30, 172, 173, 178
  see also solar photons
photosynthesis, 119
physics, 13
Pickering, Edward C., 138, 208, 210
pions, 21, 22
Planck, Max, 16, 212
Planck Satellite, 157
Planck scales, 29
planetesimals, 58, 114

planets
atmospheric gases of, 112–118
cratering on, 115, 116, 117
Earth-like, 167
orbits of, 112
plants, altering DNA of, 98
plate tectonics, 101
Pluto, 104, 168
Poincaré, Jules Henri, 121
polymerases, 75
polymerization, 55, 60
polymers, 54, 55
Ponnamperuma, Cyril, 179
positrons, 17, 172, 174
Pouchet, F. A., 42
prediction
and initial conditions, 120–121
and scientific method, 4–8, 11
presumptions, 8
Project Ozma, 184
prokaryotes, 49, 50, 79
protein binding, 74–75
protein enzyme catalysts, 63
protein enzymes, 73–74, 75, 78
Protein-first hypothesis, 56, 64
protein folding, 96, 163, 192–193
protein molecules, 94, 163, 185
and cell functioning, 71, 72
and mapping the proteome, 95–97
proteinoids, 64
proteins, 54, 55, 190–193
and cell function, 51–52
and chirality, 65
and codons, 188–190
and genetic technologies, 194
and RNA, 62
protein structure, 191–192
protein synthesis, 62
proteome, 95–97
proteomics, 95–97, 99
protocells, 62, 63
proton decay, 162
protons, 14, 15, 16, 27
in early universe, 38
and quarks, 22–23
Ptolemy, 208, 209
pulsar, 10

Q

quanta, 16
quantum chromodynamics, 24
quantum electrodynamics, 23
quantum gravitational theory, 33
Quantum Mechanical Model of the Atom, 16
quantum mechanics, 16, 17
quantum theory, 23
quark nuggets, 150
quarks, 14, 15, 22–23, 25, 26, 27
quintessence, 156

R

Rabi, I. I., 18
recycling, in scientific method, 4, 5, 7
Redi, Francesco, 42
redshift, 144
Reines, Frederick, 19
Relativity, General Theory of, 33, 134–135
Relativity, Special Theory of, 134, 212
renormalization, 30
replication (of experiments), 5–6
restriction enzymes, 85–86, 98
ribose, 59, 60, 61
ribosome, 75, 188
ribozymes, 55
rice genome, 96
Richardson, Lewis Fry, 121–122
RNA, 50–56
evolution of, 59–61
and protein enzymes, 78
and proteins, 62
replication of, 62
see also messenger RNA; transfer RNA
RNA polymerase, 75–77
RNA World hypothesis, 55–56, 61–63
Rosse, Lord, 210
Rubin, Gerry, 82
Rubin, Vera, 148
Rutherford, Ernest, 7, 15
Rutherford Solar System Model of the Atom, 7

S

Saccharomyces cerevisae, see brewer's yeast
Sagan, Carl, 65, 129
Sanger, Fred, 86
Sanger dideoxy chain termination method, 86–88
Saturn, 104
Schimmel, Paul, 90
Schramm, David, 29
Schwinger, Julian, 23
science, and technology, 1–3
scientific method, 4–11
Segré, Emilio, 172
SERENDIP project, 184
SETI (Search for Extraterrestrial Intelligence), 67, 184
Shapley, Harlow, 137, 140–141
sickle cell anemia, 192
sigma particle, 21
silicon-based life, 180
Sitter, Willem de, 136
61 Cygni, 130
Slipher, Vesto M., 144–145
Sloan Digital Sky Survey, 157
Small Magellanic Cloud, 139, 140
Smith, Hamilton O., 86, 91–92
SNAP (SuperNova/Acceleration Probe), 157
solar photons, 115
solar wind, 114, 115
Solovine, Maurice, 134
space, curvature of, 155
spaceships, and anti-matter drive, 173
Spacewatch program, 169
special creation, 40
Special Theory of Relativity, 8, 16, 134, 212
spectroheliograph, 136
spin, electron, 177–178
spin quantum number, 177–178
spintessence, 156
spontaneous generation, 41–43, 67
Standard Model (of particle physics), 6, 15, 24–28, 30, 32, 35
  testing of, 28–29
Stanford Linear Accelerator, 19, 23, 175
star catalogs, 207–210
star clusters, ages of, 168
Stardust spacecraft, 67
statistical mechanics, 99
Steinberg, Wallace, 91
stellar distances, 129–131, 140
stem cells, 80, 98
stereoisomers, 65, 66
Stoney, George, 7
strange attractor, 122, 123
string theory, 34
strong interaction, 23, 25
strong nuclear force, 18–19
sublimation, and atmospheres, 114
sugars, 65, 66
  see also glucose
Sulston, John, 81–82
Sun, 17, 58
Sun SPARCenter 2000 computer, 91
Superconducting Super Collider, 32
superconductivity, 162
Supernova Cosmology Project, 153
supernovae, 153, 157
superpartners (of particles), 34
superstring theory, 34
supersymmetry theory (SUSY), 34
Swedenborg, Emanuel, 131
synchrotron, 21, 175–176
Syngenta, 96

T

tau neutrino, 25, 26, 27
tau particle, 25, 26, 27
technicolor theory, 34
technology, and science, 1–3
telescopes, 136–137
telomerase, 195
telomeres, 195
terminal velocity, 85
Tevatron, 31, 176
theory of everything, 33–34, 161
thermal escape (of atmospheric gases), 114–117
Thomson, J. J., 7
Thomson, William, Lord Kelvin, 43
Thomson Plum Pudding Model of the Atom, 7

threose, 61
threose nucleic acid (TNA), 61
time travel, 161–162
Tiselius, Arne, 84, 85
Tomonaga, Sin-Itiro, 23
transacetylase, 74
transfer RNA (tRNA), 188
triple alpha process, 56
Turner, Michael, 154
twistor theory, 34–35
2df Galaxy Redshift Survey, 157
Twort, Frederick, 85

U

Uhlenbeck, George, 177
unification theory, 23–24
universe
  age of, 168–169
  beginning of, 29
  expansion of, 127–158
  future of, 151–152
  mass/energy contents of, 128–129
  number of dimensions in, 161
universes, multiple, 157, 169
Uranus, 104, 209

V

van Maanen, Adrien, 140
Venera spacecraft, 106
Venter, J. Craig, 84, 89–95, 196
Venus, 104–107, 118
  atmosphere of, 104, 105–106, 111–116, 118
  escape velocity of, 115
  water on, 117, 118
Venus Express, 107
viroids, 45
virtual gluon exchange, 27
virtual photon exchange, 27
virtual photons, 23
volcanic eruptions, predicting, 166
volcanoes, 114

W

Watson, James, 47, 83, 88, 90–91, 97, 191
Watson, William, 209
Watson-Crick base pairs, 50
W bosons, 27, 30, 178
weak interaction, 28
  see also electroweak interaction
weak nuclear force, 18–19
weather, 120–124
  see also climate
weather forecasting, 101–126
Weinberg, Steven, 23
Wells, H. G., 180
wheat genome, 96
Whirlpool (M100) galaxy, 132
white dwarfs, 150, 153, 169
Wickramasinghe, N. C., 66, 67
Wideröe, Rolf, 175
Wilkinson Microwave Anisotropy Probe (WMAP), 169
William of Ockham, 9
WIMPs (Weakly Interacting Massive Particles), 150
Witten, Edward, 34
Wolfram, Steven, 126, 165, 204
W particles, 23
Wright, Thomas, 131
Wu, C. S., 174

X

X-ray radiation, 9–10

Y

Yorke, James, 123

Z

Z boson, 27, 30, 172, 178
Z^0 particle, 23
Zeldovich, Iacov, 29
Zweig, George, 22
Zwicky, Fritz, 146–148, 149